The
Secrets
of Bees

The
Secrets
of Bees

An insider's guide to the life of honeybees

Michael Weiler

Floris
Books

Translated by David Heaf

First published in German as *Der Mensch und die Bienen –
Betrachtungen zu den Lebensäußerungen des Bien* by
Verlag Lebendige Erde, Darmstadt
First published in English in 2006 as
Bees & Honey from Flower to Jar by Floris Books
This revised edition published in 2019

© 2003 Verlag Lebendige Erde, Darmstadt
English text © Floris Books 2006

 Also available as an eBook

British Library CIP Data available
ISBN 978-178250-580-8
Printed in Great Britain
by TJ International

Contents

In contemplating Nature's being,
Know the One as many, seeing
In and outer coinciding,
Nothing in from out dividing.
Open secret, revelation!
Grasp it without hesitation.

Free of seeming truth's confusion,
Revel in the serious game!
Separateness is the illusion –
One and many are the same.

Johann Wolfgang von Goethe,
English version by Aldyth Morris[1]

Written in the rain

Whoever would become like a bee
who feels the sun
even through a cloudy sky,
who finds her way to a flower
and never gets lost,
to him the fields would lie in eternal radiance;
however short his life
he would rarely ever
complain.

Hilde Domin, from Nur eine Rose als Stütze

Foreword

In a time when bee-keeping has become bee-losing, when the worldwide decline of the honeybee threatens global food production, the clear voice of Michael Weiler is a much needed addition to the growing choir of those concerned about bees. His book *The Secrets of Bees: An Insider's Guide to the Life of Honeybees* is an enlightening introduction for anybody interested in bees. At the same time it is a welcome resource for aspiring and seasoned beekeepers.

The author speaks with authority because he speaks from experience based on years of devoted observation. Goethe once famously remarked: 'In vain do we try to describe a man's character, but put his acts and deeds together and the character will reveal itself.' He means that accurate observation of the facts will in time and of itself provide insights. To Michael Weiler this kind of observation has become second nature and he instructs us through his own example. Whenever he refers to graphs and statistics, it is in support of his detailed observations and not instead of them. We feel at all times carefully led and thoroughly informed by a presentation that takes us on a year-long journey.

Starting in spring, we become apprentices to a bee-master who patiently guides our gaze, sharpens our perception, points out details and explains connections we would otherwise miss. Led by his expertise we become familiar with the biography of a beehive, we witness the stages of catching the swarm, housing it, caring for and maintaining the hive. We are introduced to the miraculous reproductive capacity of the queen and follow the life-phases of the individual bee from egg to larva to worker. As a culmination our guide opens the door into the alchemical laboratory of nature. There we are initiated into the way bees transform transluscent nectar into the liquid gold that is honey. In all this we feel safe and satisfied because we are experiencing beekeeping as it should be.

At all times we have the impression that Michael Weiler's bee-craft has been learned from the bees themselves; that the beekeeping he describes is not imposed upon the hive but developed in collaboration with it. Most importantly, the symbiosis between beekeeper and bees he develops here is indeed a model for a new and mutually supportive relationship between humanity and *Apis mellifera*.

Horst Kornberger

Figure 1. Bee gathering nectar on snowberry *(Symphoricarpos)*

I. OBSERVING BEES

1. Foraging

It is a mild, dry, sunny spring day. A fruit tree stands in flower, a light breeze gently wafting its scent over us. From among the leaves a loud buzzing draws our attention. A large, flying insect is hurrying in wide curves from one flower to the next. After a while it flies away. At this time of year, it must have been a queen bumblebee. A more gentle humming persists, however, and on looking closer we see a lot of smaller bees active in the tree in the same way. All the while some are leaving whilst others are arriving, usually from the same direction. The bees busy themselves with the flowers, trying to get something from each one.

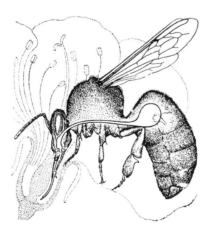

Figure 2. Worker bee showing the honey crop or honey stomach

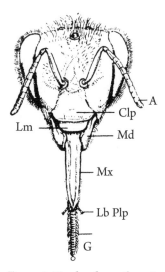

Figure 3. Head and mouth parts
of worker bee:

A = antenna
Clp = clypeus
Lm = upper lip (labrum)
Md = upper mandible
Mx = lower mandible (maxilla)
Lb Plp = labial palp (tongue feeler)
G = tongue (glossa)

Figure 4. Tip of tongue (glossa)
showing 'spoon' (flabellum)

A long, narrow, mobile organ emerges from between two
small, pincer-like jaws at the front of their heads and plunges
deep inside a flower. When seen enlarged this shows a complex
proboscis-like tongue that ends in the shape of a little spoon,
which the bee uses to take up the nectar provided by the flower.
At the same time other bees are repeatedly dusting themselves
with flower pollen. It stands out brightly against the short hair
on their furry bodies. Whilst a bee is flying to the next flower,
it quickly sweeps its three pairs of legs alternately over its body,
often hovering briefly in the air. This happens so fast that it is
hardly noticeable. On each rear leg are devices like combs and
rakes, and a fold into which the pollen is brushed. Beekeepers
call these the pollen baskets.

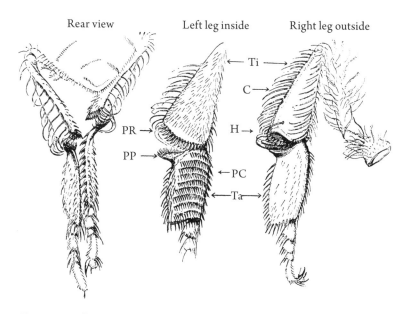

Figure 5. Pollen collection apparatus on the rear legs of worker bees

Ti = tibiae
Ta = tarsi
C = pollen basket (corbicula)
H = single hair
PR = pollen rake

PP = pollen press
PC = pollen comb
[From Zander and Weiss]

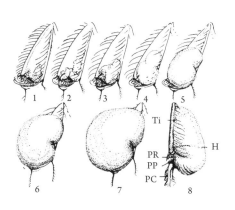

Figure 6. Forming the pollen load

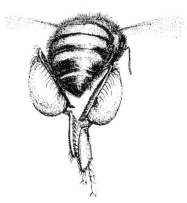

Figure 7. Bee scraping pollen into loads

Not far from the fruit tree is a partly dammed-up stream. Water glistens on the mossy roots of the bushes lining its bank as bees alight constantly on the surface of the stream. Their abdomens move rhythmically, contracting and expanding in rapid succession as though pumping. They are collecting water. After a short time they fly away again.

Nearby is a chestnut tree on whose drooping branches there are still some fat buds with a resinous gleam. The bees are busy at work there too. They are biting off pieces of the bud resin with their mandibles and transferring them with rapid, skilful movements as little droplets on their rear legs, to the place where the pollen gatherers store their pollen loads. This bud resin is rather sticky and it is amazing how the bees manage to avoid getting sticky themselves.

On a foraging flight a honey bee flies at a height of between 2–8 metres (6 ½–26 ft) above the ground and, depending on terrain and wind conditions, at a speed of about 6–8 m/sec (14–18 mph). It carries 0.05 to 0.07 g of nectar, about half its body weight, in its honey crop (hereafter referred to simply as 'crop'), which is a special compartment or sac of its anterior alimentary tract.

Bees flying continually to the same foraging site, such as a fruit tree in blossom, create a 'flight path' in the air. They make their way along this in a gently oscillating flight to their goal, never in a straight line. The unladen bees fly somewhat higher and faster, whilst those returning with nectar and pollen fly somewhat lower and more slowly.

With a lot of patience, keen eyes and a bit of luck, it is possible to discover where they are heading to or coming from. Usually this will be an apiary or a bee house, perhaps on a woodland margin or in a small sheltered clearing, or in somebody's garden. As we approach such places the number of bees in flight increases. The humming noticeable before is now louder. It is at a constant pitch which has been measured as 235 Hertz.[2]

Bees coming in from all directions impatiently head towards the hive entrances where, for the time being, they disappear from sight. Others emerge in their place and spiral up into the air, or set off immediately to an unknown destination. This constant flying back and forth, though at times sporadic, has a busy, rhythmic purposefulness about it.

We can summarise our observations thus far as follows:
- the activity of the bees at the flowering fruit tree;
- the deliberate flight movements: flying out from a central point and returning to it;
- the gathering of various things in the surroundings of the apiary;
- discovering (with the help of further investigation or appropriate reading) that the insects under observation are physically and organically adapted for their activity.

Figure 8. Bee drinking

From these observations we are able to form a mobile picture that describes a movement proceeding outwards from a central point to an invisible periphery and back again. This movement is constantly oscillating, almost like breathing as it expands and draws back in again. The sphere of activity here embraced is in constant flux: larger, smaller, now tending more one way, now switching to the other, depending on how attractive the goal is that draws the force living at its centre.

In this way we get an impression of something limb-like reaching out beyond the hive, over and over again, in order to accomplish something in its environment and bring it back.

2. In-Flight Observations

The orientation flight of young bees

With repeated visits to the apiary we can extend our observations of different kinds of bee flight. From Easter onwards on dry, warm mornings around eleven o'clock, and again in the afternoons at around four o'clock, large numbers of bees can be seen in front of the hives behaving in a totally different manner from those engaged so purposefully in gathering.

Lots of bees come out of the entrance, rise slowly into the air in front of the hive and, while flying, turn their heads towards the front wall of the hive immediately above the entrance. Then they fly up and down, back and forth, landing repeatedly on the landing board or the front of the hive, and taking off again soon after. The radii of their flights get bigger as the bees fly around the immediate vicinity of the apiary in broad, oscillating movements, like lemniscates or figures-of-eight. They soon return and disappear into the hive entrance.

Beekeepers call this phenomenon the orientation flight, or occasionally the play or practice flight. Young bees, still completing their development as hive or house bees, fly like this in the environs of their hive. The constant to and fro of the foraging flights continues unaffected.

Figure 9. Bees in flight

The flight of the drones

However, we might notice yet another surprising detail. With a loud buzz, deeper in pitch than the general hum of the apiary, a large insect approaches fast and disappears into a hive entrance. From other hives, especially those with heavy flight traffic, similar insects emerge, move heavily across the landing board and take off with a loud drone. At first, they rise somewhat clumsily into the air, almost staggering, but then they ascend with a powerfully direct flight and disappear into the environs. Despite their size and their droning flight sound, it is difficult to keep them in sight for long because they fly away so quickly. These fliers are called 'drones' because of the noise they make.

Drones are very different from foragers. They are longer and fatter. Their compound eyes are much bigger and cover almost the whole of their heads. The mobile antennae that grow from the 'middle' of the 'face' have one segment more than those of foragers: eleven instead of ten. The mouth parts of the drones are also less developed than those of the workers: the proboscis is significantly shorter and less differentiated, and the mandibles are smaller and shorter. The furry thorax of drones is fatter and the pairs of wings are so big that they almost extend beyond the end of the abdomen. The three pairs of legs seem barely capable of supporting the body and they are less differentiated in their outer segments. They also lack the forager's pollen-gathering apparatus of rakes and combs and baskets. The abdomen is longer too with the effect that drones are about half as long again as workers. They possess neither sting nor poison gland. Instead, with suitable pressure on the abdomen, a milky-white strangely shaped organ emerges that biologists call the male sex-organ.

We might think it curious that we have never seen a drone before in the open. We see them neither on flowers nor inside our house windows, where sometimes a worker bee out foraging is trapped after taking a wrong turn. And yet it is

Figure 10. Drone

a fact that they leave the apiary and come back again. But where do they go and where do they stay during their flights that last on average from 45 minutes to an hour? We shall come back to this question later.

A hive swarms

Now another event holds us spellbound. It is close to noon. The sun is almost at its zenith. For some time there has been a certain amount of disturbance at one of the hives. Lots of drones have been flying in and out. The orientation flight has just finished, but quite a lot of bees are scurrying around the hive entrance, going in and out or rocking backwards and forwards as if scrubbing (this is known as the 'washboard movement'). The busy foraging flight is replaced by a new activity. Masses of bees are pouring out of the hive entrance almost like a stream of water, tripping over and bumping into one another and immediately rising into the air, which is soon filled with a loud buzzing. While the outflow has still not finished, the bees that have taken off fly high into the sky. They do not head purposefully to some point like the foragers, but instead fly around in large figure-of-eight loops near to, or further away from, the apiary: some quickly, others more slowly and peacefully. It seems as if no spot in their entire surroundings is not visited or flown through by a bee. They alight on the grass and on the bushes and the trees; they crawl around and take off again. They even approach the astonished onlookers and crawl around on them as if looking for something.

Meanwhile the flow from the hive entrance has dried up. Instead a small collection of bees has gathered on a tree a few metres from the hive and more are drawing to it. If we approach cautiously we may even notice that one bee in particular, longer and more slender that then rest and with gleaming red legs, is being covered by the new arrivals. By now a cluster of bees about the size of a pear is hanging on the branch. On the outside of it are quite a number of bees with their abdomens pointing upwards into the air, their tips bent downwards somewhat. This exposes a light brown, slightly everted area between the penultimate tergite segments. The bees are fanning their wings powerfully, producing a noticeable flow of air. This light brown area is the Nasonov scent gland and the bees with their bottoms in the air are fanning out a current of scent. This is known as 'scenting' and it attracts the bees still flying around to the cluster.

It is now at least 5 minutes since swarming began. The hum of the swarming bees is creating a uniform sound picture. The number of bees flying to the swarm cluster is increasing and it is slowly getting bigger. More bees join in the scenting, and the flight paths of those still in the air are shortening. About 15 minutes from the start almost all of the bees are in the swarm cluster, which has now grown to roughly the size

Figure 11. Bees fanning with their scent glands exposed

Figure 12. Swarm cluster

of a rugby ball. A swarm of this size weighs around 2.5 kg (5.5 lbs) and contains about 18,000–20,000 bees along with the stores of honey they consumed before swarming. There are also drones in the swarm and bees with pollen baskets which have joined the swarm after returning from their foraging flight. After twenty minutes, peace returns once more to the apiary; bee traffic is as before. From a short distance away, the swarm appears to hang motionless in the tree. Most likely a casual passer-by would not even notice it.

We will look at the swarm and what happens to it next in the following chapter, but first let us return to the hive.

The flight of the Queen

Ten to twelve days after the hive threw the swarm (beekeepers say that a colony from which a swarm has issued has 'thrown' a swarm) everything about it looks the same as the other hives in the apiary. The orientation flight of the young bees has just

Figure 13. Queen bee

finished; drone flights have increased. After a while, when the sun has just passed its zenith, a bee appears at the hive entrance that looks different from the familiar workers and drones. She is slender and longer than the others, her abdomen extends well beyond her wings, and her legs are somewhat longer with a reddish gleam. She immediately takes off. Due to her distinctive size it is easy to distinguish this bee in flight from the others. She flies in a few arcs in front of the hive, climbs higher and disappears quickly in a specific direction. We may note that there are no other bees of this kind in the colony.

After about half an hour has passed this large bee reappears. The bees on the landing board come up to her and touch her with their feelers and forelegs and stroke her with their proboscises. A white appendage is hanging out of the tip of her abdomen, gleaming somewhat moistly like a little pennant. She walks in through the entrance and disappears. This is a queen bee and she has just returned from her mating flight.

Drone congregation areas

So far nobody has been able to observe the open-air mating flight of a queen bee, but conditions that give an idea of this event can be arranged experimentally.

For a long time, it was not known why there were drones in a bee colony, and people did not know where they flew to. Reports of so-called 'drone congregation areas', supposedly somewhere in the vicinity, were always cropping up without anyone knowing exactly what they were. More recently it has become possible to ascertain certain facts about these places.

Every year, from the end of April to the end of August, drones from neighbouring colonies congregate in a specific area from around eleven o'clock in the morning until around five o'clock in the afternoon. They fly back and forth so fast that they can be hard to spot, and these places are usually only recognised by the low hum of the drones. If a lure is made with a caged queen fixed to a helium balloon attached to a string, the height and extent of such areas can be determined. As soon as she is pulled out of the congregation area, no more drones will follow. In such an artificially contrived way it has been possible to film the mating of a queen.

From these observations it has been established that in her mating flight the queen flies to a drone congregation area and is there chased by the mature drones and mated. The drone that reaches the queen clasps her abdomen from above with his hind legs. By squeezing all the air and fluids in his body into his abdomen he everts his genitalia. During the ensuing copulation this is firmly locked into the opening of the queen's genitalia at the tip of her abdomen and this injects the seminal fluid into the queen's spermatheca. The drone then tips over backwards, his genitalia are torn out and remain hanging from the genital orifice of the queen. The drone falls dead to the ground.

Mating can happen with several drones in succession, on average between five and fifteen. Each subsequent drone removes the 'mating sign' of its predecessor. The queen brings the last sign back to the hive.

3. The Swarm

Twenty minutes after the swarm has left the hive, calm has returned to the apiary. The mood of excitement that filled the whole area with thousands upon thousands of swarming bees has once again returned to a constant hum of departing and returning foragers, modified only by the droning buzz of the drones. The bees have now gathered at a single point. After the out-swarming comes the concentration: diastole and systole on a grand scale.

The swarm cluster

The swarm is hanging motionless on a branch in the shadow of the crown of the tree and is not immediately obvious. A beekeeper who was not there to observe the hive swarming would have to spend some time looking for it, if the cluster was not hanging at a place where swarms had frequently landed in the past. As a matter of routine at swarming time, beekeepers scan the trees and bushes around the apiary as soon as they arrive, but they can also discover swarms because they notice a certain 'irregularity' which is introduced into the usual order of the surrounding vegetation.

But beekeepers can also have a clear overall view of a swarm, and for the time being we shall take a look at one of these.

After a short time, we may notice, on the outside of the

swarm cluster, a few bees who are behaving in a particular way. A bee walks over the cluster for a short distance in a straight line while at the same time rapidly waggling her abdomen from side to side. Then she returns in a semicircle without waggling her tail. Back at the starting point she repeats the waggle along the short straight line and this time returns to the starting point in a semicircle in the opposite direction to the first. She repeats this for several minutes, each time with the semi-circular path alternating right and left of the straight line (making a figure of eight, see also Figure 34, page 73). After watching this for some time it becomes clear that the straight line always points in the same direction and the walk maintains a precise rhythm, for example, five circuits in fifteen seconds.

Longer study reveals that other bees show the same behaviour only with different rhythms and directions of the 'wagtail' walk. Whereas the waggle dance of one bee always points upwards, another always points almost horizontally and the frequency of its circuits is lower. A third bee is observed pointing downwards to the left and making more frequent circuits than the others. Her waggle seems more vigorous. Other bees go up to this one and touch her with their antennae

Figure 14. Swarm cluster

and briefly follow her after which they disappear into the cluster or fly away. More will be said about this bee behaviour, also called the 'waggle dance', in Chapter 7, where we will consider the waggle dance as an inner expression of the relationship of a bee colony to its surroundings.

Busy flight traffic gradually develops from the suspended swarm as several bees take off from its surface, whereas others use a hole that opens in it at the bottom end forming a kind of hive entrance. The cluster contracts and is less active at lower temperatures.

If it rains, the bees on the outside arrange themselves in layers like roof tiles with their heads facing upwards and their wings and abdomens downwards and outwards. It can rain for a long time without water penetrating the inside of the cluster.

If we watch the dance behaviour over a longer period, perhaps for several days, we will notice other features. Gradually an increasing number of dancing bees on different parts of

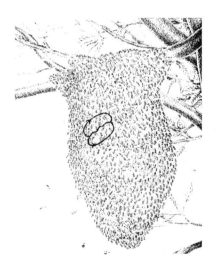

Figure 15. Swarm of bees, illustrating the pattern of the waggle dance

the swarm-cluster point in the same direction, for example, in the third direction described above. Several dance in the first direction too, and the second has disappeared completely. Other directions of movement are also observed repeatedly but they eventually disappear.

One particular direction of the waggle dance begins to predominate, possibly after 2 days, and is then performed in the same way by the majority of the dancing bees. Suddenly the behaviour of the swarm changes. Vibrating their wings, the bees run around very fast seemingly in a muddle. Things start to loosen up, the cluster starts to disperse. In a short time, the bees leave the place they started from and fly away, all in the same direction, in an elongated cloud. If they do not fly too far, the observer may be able to follow them and can watch how they gather at a new place and there disappear, perhaps through a hole or a crack, into a cavity. This could be a hollow tree or a crack in a rock; but it could also be the false ceiling of a half-timbered house or a cavity wall. Other possibilities include large empty nest-boxes, chimneys and, not least, an unoccupied hive left open at another apiary.

At the opening of this cavity a pattern of flight activity develops that is similar to that already described for the hive in the apiary. What is going on in the cavity is out of sight and for the time being can only be imagined.

Catching a swarm

Only if a beekeeper 'takes' a swarm is it possible to follow its subsequent development. For this an empty container is needed, which could be a cardboard box, a tub, or anything similar that is bee-tight, closable, and allows sufficient ventilation. It is also useful to have a feather duster (for example, a goose wing) or a bee brush. It also helps to spray the outside of the cluster with a little water as this causes it to contract.

The box is held under the quietly suspended bees and the branch is given a vigorous downwards shake. The cluster drops into the box like a ripe fruit and there it flows around like thick soup. The few bees still hanging onto the branch are brushed in with the rest and the box is then closed.

Once it is certain that almost all the bees are in the box then it can be taken away. But if a lot of bees take off or miss the box when the swarm is knocked down then they need to be coaxed back. The beekeeper makes a flight entrance for the bees in the box and places it on the ground nearby, preferably in a shady place and ensuring if possible that any breeze there may be is blowing from the box to the place where the swarm was hanging. The bees in the box are starting to roar and fan scent again; those still outside are gathering at the starting point on the branch or flying around in the familiar lemniscatory loops. While they do this, they are able to detect the scent and soon find their way to the swarm in the box.

Figure 16. Flight entrance of a wild colony in a tree with a curtain of propolis in front (Entrance diameter: 6 cm/2 in)

Figure 17. Swarm catching box with the swarm that has been knocked into it.

By that evening at the latest, the swarm that has settled in the box can be taken away. The beekeeper places it in a cool, dark, quiet place, for instance a cellar. In this situation the bees will hang from the top of the box thus re-forming the swarm cluster. This further development can easily be observed in a purpose-built swarm-catching box, fitted on two opposite sides with closable ventilator grilles. When the two opposite sides are uncovered carefully a little at a time and some light is provided in the background, it is possible to see the silhouette of the mass of bees hanging like a bunch of grapes or in a heart shape.

The swarm in the box is very quiet, with the bees giving off a brief, loud wave of hissing-humming only if they are disturbed.

The beekeeper may leave this box undisturbed for up to three days. The stores that the bees have brought with them will only last that long. A swarm left alone undisturbed like this will lose between 100–200 g (3 ½ –7 oz) in weight during this period. Alternatively, a swarm may be hived the evening after it is taken.

Hiving the bees in a new home

In the meantime, the beekeeper prepares a new hive for the swarm. The brood box is cleaned and equipped with frames for supporting comb. These are made of strips of wood, usually referred to as 'bars', which form a rectangular shape, longer in either the vertical or horizontal dimension depending on the type of hive being used. They are hung vertically and at equal distances from one another, then the finished hive is closed up with the hive entrance left open.

In the late afternoon of the hiving day the beekeeper collects the swarm from the cellar. The box is opened carefully. The swarm cluster is still hanging quietly inside. At the entrance to the new hive the beekeeper has placed a horizontal board onto which the swarm is now briskly shaken. The mass of bees flows

Figure 18. Swarm cluster in the box just before shaking it out

Figure 19. Swarm being hived

over itself like dough. A few bees immediately rise into the air again and fly around the area in lemniscatory curves, many start scenting in the way described earlier, but most of them creep in a uniform flow towards the hive entrance and steadily disappear through it into the hive. Once again, the queen can be spotted as she is driven towards the hive entrance in the general flow.

Other bees slowly take off, many start scenting in the way described above and a subtle, difficult-to-describe, scent is noticeable in the immediate vicinity.

Meanwhile the beekeeper has carefully brushed the last few bees out of the swarm box. Three white structures are hanging parallel to each other the same distance apart on the upper wall of the box. They are heart-shaped, oval and about 2 cm (1 in) thick: the one in the middle is about twice as big as those either side. Closer examination reveals that the swarm has made three beautiful white combs with the characteristic hexagonal cells. Small white rods about 2 mm (⅛ in) long can be made out on the bottoms of some of the cells. These are eggs laid there by the queen.

On the bottom of the swarm box are a few dead bees, some small twigs, and a large number of white, oval, opalescent platelets, between 1–2 mm (¹⁄₁₆ in) wide. These are scales of beeswax which the bees have 'sweated' out and which have fallen down during the construction of the comb.

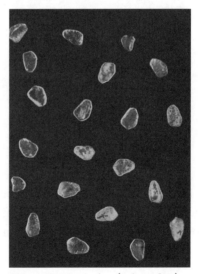

Figure 20. Wax scales, (enlarged 4x)

Figure 21. Comb with bee

The spacing between the combs in the swarm box is 35 mm (1 ½ in) and the frames in the hive have been hung accordingly so that the colony will build comb in them. This allows the combs to be moved individually later on by the beekeeper.

After about fifteen minutes, most of the bees have disappeared into the hive, the last few gently brushed off the hiving board in front of the hive entrance.

II. INSIDE THE HIVE

4. Building Comb

The following bee activities take place inside the hive. These processes can only be observed by the beekeeper by opening the hive from time to time. The movability of the frames allows detailed observations to be made, and the accumulated scientific research in the specialist literature enlarges the picture and gives us a fair idea of the life of bees inside the hive.

To study the hexagonal cell structure, a single comb is lifted out of the living whole. However, the comb itself can only be seen if the bees have left it or been shaken or brushed off it.

Figure 22. Comb on a frame

▲▼   Figure 23 and 24. 'Alfresco'

How is comb created? Our first impressions are from the comb made by the swarm in the swarm box, and from the small round waxy platelets 'sweated' by the bees. What is the explanation for this process?

On the underside of the worker bees' abdomen are glands arranged in four pairs from which the bees can secrete wax. As they cannot manage this alone, certain conditions are necessary for them to do this.

The conditions for wax production are:
- a sufficiently large colony with a queen;
- an adequate food supply with carbohydrates (nectar, sugar) and protein (pollen);
- a cycle of increasing development (build up), or a swarm situation;
- enough bees with activated wax glands.

Shaping space

The bees link together in a sheet by forming chains and hang from the lid or some other upper boundary (top bar of a comb frame etc.) in the cavity that they occupy. Increased heat production can be detected in the construction cluster (greater than 36°C or 97°F) as a number of bees exude wax scales. These are carried upwards with the legs and mandibles. The bees at the top of the construction cluster knead the wax with their mandibles while at the same time probably mixing it with a secretion from one of their cephalic or head glands (see Figure 33 on page 66). Then they attach it to the frame bar from below. The wax is first drawn into blobs and a little later is shaped into a hexagonal matrix. By applying more wax to the edges of the developing comb its area is increased. As soon as this has reached a diameter such that, if rotated around its vertical axis, it would sweep out a sphere of about 5–7 cm (2–3 in), they start the construction of adjacent combs. This way the comb matrix grows until eventually it hangs in the space occupied by the swarm.

The development of the process can be experienced as

a movement. The swarm cluster, or colony, permeates a particular volume of the cavity in which they have made their home. Now the wax glands of a number of the bees are activated and wax is produced and exuded. Individual bees can manage this because they are part of a colony. The wax is taken up by the suspended cluster and moulded into its characteristic comb formation in an interior space.

Comb building stops when one of the above-mentioned conditions is no longer fulfilled, such as when the comb as a whole fills the available space in the cavity, or when its size reaches the limit of what can be occupied and permeated by the formative capacities of the colony.

Figure 25. Natural comb construction in a rear-access hive (a Baden three-storey hive) set up 'warm way' by the swarm.

But what happens to a swarm that has found neither a suitable cavity nor has been taken by a beekeeper? In such an emergency it builds its comb out in the open, for example on the branch of the tree on which it has settled.

Beehives

In order to continue a more detailed study of the processes inside a bee colony, the hive has to be opened for a little while every day. All the frames that already contain constructed comb have to be lifted out of the box.

There are two basic types of hive: those that open at the top and those that open at the back. The frames may be shorter than they are wide (broad format), or they may be taller than they are wide (tall format), although the latter is rare in the UK and USA. Frames are rarely, if ever, square.

The next distinction concerns whether the frames are hung parallel to the front of the hive or at right angles to it. The former is referred to as the 'warm way', the latter as the 'cold way'. This terminology is connected with opinions on air flow in the beehive.

This arrangement of frames produces clear differences in the construction of the comb. With the warm way the swarm build in the front frames hanging down parallel to the front wall of the hive. In doing so, the second and third combs soon become the biggest, with the comb at the front remaining smaller at first. The frames further back are gradually built in too, being started later. In warm way hives with rear access, the rear closure is usually a window which allows these developments to be observed without opening the hive completely. With 'cold way' the swarm usually occupies the frames in the middle and from there builds the layers of comb to each side frame by frame. In this case the central comb is at first the biggest whilst those at the sides start off smaller. Only when the colony has built up sufficient strength to permeate the entire space do all frames gradually get filled with comb. Depending on the size of the swarm, this can take several weeks or, if it is being started at the beginning of summer, it might not be completed at all that season. Most of the job is completed in the first 3 weeks, but construction of the comb can last for several weeks after hiving the swarm.

5. Bee Development

The following description is based on inspecting a swarm in a top-opening hive. The frames are hanging cold way and are tall format.

After the beekeeper has removed the hive roof, at least one more cover comes into view. It may be a wooden 'crown board' or a waxed cloth known as the 'quilt' or a transparent sheet. Beneath this a gentle activity can be detected. If we place a hand on this surface it will feel warm in varying degrees and in different places; it is warmer where most of the bees appear to be. The beekeeper slowly lifts up the sheet. A few bees come up on to the top bars of the frames and may fly away. A soft roaring increases from amid the general humming. The beekeeper gives a few puffs of smoke over the frames with his smoker, whereupon some of the bees go back down amongst the frames. 'Where there's smoke there's fire,' and animals generally flee from fire. Honey gatherers have been putting it to use since ancient times.

It is now the third day since the swarm has been hived and it is clear that the frames to the right and left are still free of comb. The swarm has suspended itself in the middle of about seven or eight frames. The free frames are removed and put to one side. Looking diagonally into the box from above, we can see that the dark mass of bees are suspended just as they were in the swarm-catching box. With the removal of the next frame and the one after that, the bee cluster is split apart. Here a certain resistance is felt, demonstrating how tightly the bees are holding on to each other and to the wood of the frame. The beginnings of a comb are visible in the frame next to the central one. In the middle frame there is a comb about the size of a hand, and on the frame after it there are also the beginnings of another comb.

By lifting the frames carefully, lots of bees can be seen hanging in chains on the lower margin of the comb. There are several vertical chains as well as some going diagonally or even almost horizontally to the side bars of the frame. Each bee clings to the next with its forelegs and hind legs like links in a chain. If we try to break such a chain with a stick or a finger there is a significant resistance to be overcome. However, almost all are broken when a frame is lifted out and most of the bees fall into the box.

Egg laying

In some of the cells at the upper edge of the comb is a thick, gleaming, transparent yellowish fluid. This is either the honey reserves that the bees brought with them, or freshly gathered nectar.

On both sides, near the middle of the comb, are cells containing eggs. These patches, about 4 cm (2 in) wide, comprise about fifty cells each, at the bottom of which are fixed little white rods.

At the next inspection 2 days later, we can see that the area of comb has grown. The fourth and fifth combs have been started and the middle one is now 17 cm (7 in) tall and heart-shaped, somewhat longer than it is broad. The area of cells in which the queen has laid eggs is now at least as large as the palm of a hand and there are patches of eggs on both adjacent combs. A rough calculation shows that between 300 and 350 eggs have been laid on either side of the central comb, and on the adjacent combs there are about 150, meaning the queen has laid about 1,000 eggs since she started.

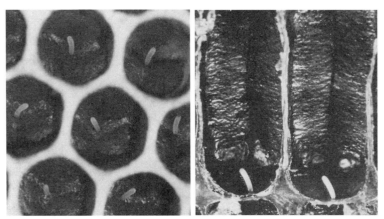

Figure 26. Eggs in cells

left: viewed from the entrance to the cell
right: cross-section view from the side

Figure 27. Laying queen surrounded by ring of nurse bees ('court')

During this inspection we notice the queen on a comb surrounded by a star-shaped ring of bees. Some are stroking her with their antennae. Others are offering their mouth parts and proboscises to her and she is taking something from them with her mouth.

The bees are constantly feeding their egg-laying 'mother' with a highly nutritious substance known as royal jelly. This is a special secretion from the hypopharyngeal glands in their head and is produced by young bees at a particular stage of their development, provided they consume sufficient quantities of honey and bee bread (see page 136). The nutritional physiological analysis of this liquid shows that it is a mixture of easily digested carbohydrates and balanced amino acid compounds, or first-class proteins.

In the upper part of the comb there are significantly more stores of honey and nectar that have been deposited there in the previous two days.

In a healthy colony between Easter and St John's Tide (24th June), the queen bee lays an egg in a comb cell roughly every 40–60 seconds. Sometimes this amounts to 2,000 cells in a 24-hour period. The accumulated weight of these eggs greatly exceeds the body weight of the queen (0.16–0.23 g). This creative impulse constantly emanates from the queen through the bee colony by her egg-laying activity, and this is matched by the steady supply of royal jelly to the queen. If this flow diminishes or is interrupted, egg production diminishes in a short time or ceases altogether.

Larvae

More discoveries are made at the next inspection a few days later. The middle comb has in the meantime grown so large that it has become attached to the side bars of the frame. In the cells in which the first eggs were laid are now small, curved, white grubs, or larvae, lying in a clear-to-whitish translucent liquid on the bottom of the cell. A liquid gleams here and there in the surrounding cells too, and on close inspection it is possible to see that the eggs here have also become tiny larvae. As a rule, about 72 hours after an egg is laid by the queen a larva hatches out of it The liquid in which the young larvae float is almost identical to royal jelly.

The area of comb laid by the queen has now grown and covers parts of the fourth and fifth layers of comb. In the past two days she has laid a further 1,000 eggs. Chains of comb-building bees have formed on the sixth and seventh frames.

On the fourth and fifth frames, which at this time form the outer combs of the developed nest, some cells have been filled with a dark substance. Many of them are deep yellow, others violet or black. In one of the cells there are two small orange clumps. This is where the pollen baskets are being taken, to be transformed into 'bee bread': a mixture of pollen and honey that has undergone a fermentation process, which makes it both digestible and preserves it for bees (See Chapter 19. Pollen, Bee Bread and Royal Jelly, page 136). Flower pollen in the form of bee bread provides the protein component of the diet of the bees and their brood.

Three days later, the beekeeper has another look at how the swarm has developed. It has been in the hive for ten days. Comb has now appeared on the sixth and seventh frames and is already the size of a hand. Honey and pollen stores have been deposited in it. On the fourth and fifth combs, there are large patches of eggs, and the eggs that were visible three days ago have become small larvae. On the second and third combs the patches of brood have increased in size. The most recent and young larvae are surrounded on each side and below by a large area with eggs. In the meantime, these combs have also been fixed to the side bars of their frames. The middle comb has grown longer at the bottom and here, too, a broad ring of eggs surrounds an area of comb with larvae whose age increases towards the middle and the top.

Where eight days previously the first eggs had been laid, there are now fat round grubs filling the diameter of their cells. These are called 'round larvae' and it looks as though they have hardly any more room to continue growing.

They swim around in their food, always rotating round their tails in the direction they are facing. The older larvae are no longer fed with royal jelly from the cephalic glands of the nurse bees (as those that feed and care for the brood are called) but with a mixture of pollen, honey and royal jelly.

Pupation

The next inspection happens three days later. Vigorous foraging flight is underway at the hive entrance. The many pollen gatherers are conspicuous.

Inside the hive, the bees have not started to build any new combs. Where still possible, the existing seven combs have been extended to the sides of the frames and in all cases to the bottom. The stores of nectar and pollen have only grown a little. Obviously

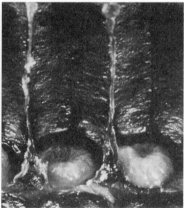

Figure 28. Round larvae

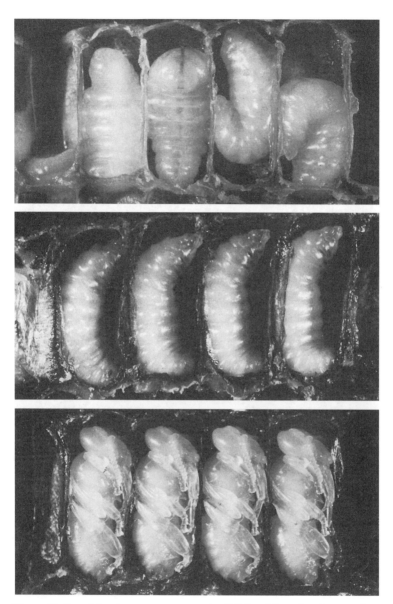

Figure 29. Brood development in a capped cell ('closed for redevelopment'):
stretched larvae spinning cocoons (top)
before pupation (middle)
pupae (bottom)

Figure 30. New comb with capped brood

Figure 31. Wild comb with worker and drone cells

much of the food is now being used for feeding the 'open' brood.

In the combs at the sides, large patches of cells with eggs have appeared and the eggs that were visible three days previously have now become the youngest larvae.

The appearance of the middle comb has changed a lot. The cells that eggs were seen in at the first inspection ten days ago are no longer open, and the cells around them have acquired waxen, slightly domed caps. Where that has not occurred larvae are still visible, although they are no longer round and fat and lying on the bottom of the cell, but instead have straightened out towards the entrance to the cell and are almost looking out of it. These more mature grubs are 'stretched larvae'. The cells are capped by the bees nine days after egg-laying, at which point a round larva of worker brood weighs about 0.15 g, more than 500 times the weight of the egg. In the case of 2,000 larvae of approximately the same age, this corresponds to a weight increase of 300 g in six days.

During the next three days the larva spins a cocoon around itself, becoming a pupa that changes into a bee. At the end of feeding, a stretched larva weighs about 0.12 g (less than $\frac{1}{100}$ oz), which is at least 500 times heavier than the egg that the queen laid nine days ago.

Drone brood

The next inspection occurs seven days later, on the twentieth day after hiving the swarm and the seventeenth day after the first inspection. The layers of comb have again grown bigger. Areas as big as a hand have now appeared in the eighth and ninth frames and the bees would probably have started building in the tenth and eleventh frames, but no chains of wax-making bees are to be seen there. The swarm builds vigorously for three weeks and then its construction productivity declines. This is dependent on the strength of the swarm and on the forage and weather conditions. Pollen has been deposited in the outer combs together with some honey. The middle three combs have now reached the bottom bars of their frames.

The condition of the fourth comb has also changed. There is a small gap between the comb and the bottom bar, and the cells in the lower region look different. They have openings that are at least a third larger in diameter than those we have seen hitherto. There are eggs at the bottom of these cells too. In these cells the drones of the bee colony are being raised. The drones are bigger than the worker bees.

The brood nest has increased in size and almost all the brood that was present at the last inspection has now been capped. Around it, up to the penultimate combs to the sides, are areas of brood with all the familiar developmental stages.

Hatching bees

Four days later the picture has changed once more. The cells that were capped in the middle of the upper part of the central comb are now open again. Compared with the shining white comb at the start when the eggs were laid by the queen, they are tinged a brownish colour. This is from the cocoons that each hatching bee has left behind in the cell, a residue of its metamorphosis. These cocoons comprise the pupal silk, larval faeces, and residues of the five moults (ecdyses) that the larva has undergone.

A few of the surrounding cells are also empty. Some bees are crawling around on them that look very different from the other bees. They look as if they are surrounded by a fine grey fur and their wings are somewhat creased and not yet unfolded. They extend their proboscises to other bees and are fed.

In one of the capped cells a small hole appears. Half a mandible pincer sticks out and begins to gnaw away at the capping against the other mandible hidden from view. As

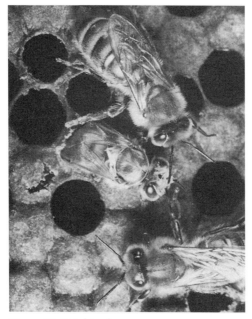

Figure 32. Hatching bees

the hole gets bigger, the pincer is withdrawn and an antenna appears briefly, feeling about on the cell capping. The gnawing resumes. The removal of the capping has proceeded further on another cell and the face of a bee looks out. The antennae are feeling around and the head is extended outwards. Leg segments come into view and, with a lot of effort, the round thorax is slowly squeezed out. Once it is out, the abdomen quickly follows. The small, crumpled bee looks somewhat lost amongst the teeming activity on the comb. Shortly afterwards, it is offered food from the proboscis of another bee and takes it immediately.

A hatching bee is only fed if it has left the cell through its own efforts. It is not helped during hatching. If a bee cannot accomplish it on its own it is killed and the corpse is removed from the hive.

The developmental stages of the *worker brood* are as follows:
- Egg: 3 days
- Young larva: 3 days, fed with royal jelly
- Older larva: 3 days, fed with honey-pollen mixture
- Cell capped: 9 days
- The round larva straightens into a stretched larva and spins a cocoon round itself up to the 12th day of development
- After 12 days in the capped cell the young bee hatches on the 21st day after egg-laying by the queen

Therefore the brood period lasts 3 + 6 + 12 = 21 days.

The cycle of development for *drones* is different from that of worker bees:
- Egg: 3 days
- Larva: 6–6 ½ days
- Cell capped: 10 days
- Round or stretched larva: 8–14 days

Drones undergo at least 14 days of metamorphosis. This means they hatch at least 24 days after egg laying, although it can be as much as 25 or 26 days. The entire sperm supply is fully developed by the round and stretched larva stages, although a drone is not mature enough to fertilise a virgin queen until 10–14 days after hatching.

The development of the queen is likewise different from drones and worker bees, and this will be presented in Chapter 8 (see Table 2 on page 80 for an overview of the developmental stages of queen, worker and drone bees).

The timings for brood development are relatively inflexible. There is a constant temperature in the brood nest of the hive of about 35°C (95°F) and an almost constant humidity of about

60%. Only when extreme circumstances outside the hive disturb the internal conditions, or if the brood is damaged or dies, might the development be increased by one or two days.

A further inspection a few days later reveals that the area of hatched cells has increased. If the colony has enough forage, a portion of the vacated cells are filled with stores. The queen lays eggs again in the rest.

In trying to take in all these life processes in a mobile overview we might have the impression that the queen's laying activity sends an impulse of rhythmic regularity throughout the colony. We shall consider later whether there is any evidence in this impulse of movement in any direction.

6. Bee Life

The queen lays eggs in a spherical space of increasing size within the layers of comb that the swarm has produced since it was hived. After a few days, and for a period of time, she lays over 1,000 eggs every 24 hours in the newly created cells. The first young bees of the new colony emerge on the 23rd day after 21 days' development. After that, more hatch each day. This is absolutely essential for the further development of the swarm. Lots of bees are lost daily during foraging activity. They fail to return to the hive from its surroundings because of age, exhaustion or other influences. As a result, the number of bees in the swarm decreases daily from the beginning. But the hive must have sufficient bees to sustain the essential life processes within it.

House bees and flying bees

For the purpose of observation, we can divide the life processes at work into an assortment of activities within the colony. They take place over a specific period and are carried out either by individual bees or in collaboration with other members of the hive. These include guarding the hive entrance, building comb, exuding wax, feeding the brood, warming the brood nest, and cleaning and clearing out. There are many more of these activities.

If we were to watch individual bees of a colony over a period of many days, we would probably notice at first that the bees can be divided into two kinds: flying bees and house bees. The latter are responsible for the maintenance of the life processes inside the hive, the former for the nectar and pollen gathering activities we have already discussed. It is possible to make this distinction because it is clear that the flying bees (also known as foraging or field bees) do not take part in the work of the hive; they simply unload what they have brought in terms of nectar, pollen and water in order to fly out again. As the number of flying bees needs to be kept relatively constant in relation to the strength of the colony, there needs to be a daily replenishment of those out hunting.

In the hive there is a great variety of jobs to be carried out, and this raises the question of how the necessary division of labour arises.

Bee researchers conduct experiments by carefully marking a single bee, or several of them, with a small coloured blob on the back of the thorax immediately after hatching. This enables them to trace the bee wherever it goes.

In order to do this, observation hives are constructed. These contain only one large layer of comb covered on both sides by sheets of glass, which enables the activities of all the bees to be monitored. In Greece, 2,500 years ago, Aristotle observed the life of bees in this way and described what he saw. Instead of glass he used selenite (crystalline gypsum, calcium sulphate). Maurice Maeterlinck also observed bees in this way and wrote about them in his book, *The Life of the Bee*, published in 1901.

Development from the inside...

A newly hatched bee is first fed via its proboscis by other bees. Its movement over the comb continues unassisted. It almost

Table 1. Bee development. Course of brood events starting from a swarm situation

Swarm age (1ˢᵗ day = hiving)	1	2	3	4	5	6	7	8	9	10	11	12	13	14	15	16	17	18	19	20	21	22	23	24	25	26	27	28	29	30	31	32	33	34	35	36	37	38	39	40	41	42
Inspections		I		II		III				IV			V						VI					VII																		
Brood age	1	2	3	4	5	6	7	8	9	10	11	12	13	14	15	16	17	18	19	20	21																					
			1	2	3	4	5	6	7	8	9	10	11	12	13	14	15	16	17	18	19	20	21																			
				1	2	3	4	5	6	7	8	9	10	11	12	13	14	15	16	17	18	19	20	21																		
					1	2	3	4	5	6	7	8	9	10	11	12	13	14	15	16	17	18	19	20	21																	
						1	2	3	4	5	6	7	8	9	10	11	12	13	14	15	16	17	18	19	20	21																
							1	2	3	4	5	6	7	8	9	10	11	12	13	14	15	16	17	18	19	20	21															
								1	2	3	4	5	6	7	8	9	10	11	12	13	14	15	16	17	18	19	20	21														
									1	2	3	4	5	6	7	8	9	10	11	12	13	14	15	16	17	18	19	20	21													
										1	2	3	4	5	6	7	8	9	10	11	12	13	14	15	16	17	18	19	20	21												
											1	2	3	4	5	6	7	8	9	10	11	12	13	14	15	16	17	18	19	20	21											
												1	2	3	4	5	6	7	8	9	10	11	12	13	14	15	16	17	18	19	20	21										
													1	2	3	4	5	6	7	8	9	10	11	12	13	14	15	16	17	18	19	20	21									
														1	2	3	4	5	6	7	8	9	10	11	12	13	14	15	16	17	18	19	20	21								
															1	2	3	4	5	6	7	8	9	10	11	12	13	14	15	16	17	18	19	20	21							
																1	2	3	4	5	6	7	8	9	10	11	12	13	14	15	16	17	18	19	20	21						
																	1	2	3	4	5	6	7	8	9	10	11	12	13	14	15	16	17	18	19	20	21					
																		1	2	3	4	5	6	7	8	9	10	11	12	13	14	15	16	17	18	19	20	21				
																			1	2	3	4	5	6	7	8	9	10	11	12	13	14	15	16	17	18	19	20	21			
																				1	2	3	4	5	6	7	8	9	10	11	12	13	14	15	16	17	18	19	20	21		
																					1	2	3	4	5	6	7	8	9	10	11	12	13	14	15	16	17	18	19	20	21

Brood stages:

1	
2	
3	Laying
4	
5	
6	open brood: 4 – 6 younger larvae (grubs)
7	7 – 9 older larvae (grubs)
8	
9	
10	capped brood: 10 – 12 stretched larva (grub)
11	
12	
13	13 – 20 pupa
14	
15	
16	
17	
18	
19	
20	
21	21 hatching

— ≫ — ≫ — ≫

staggers. Its hair appears grey, matted and ruffled. With hesitant movements at first the downy grey young bee cleans itself and smooths out its still crumpled wings. It looks like a little, helpless child, but soon seems steadier and more resolved. After several feeds it begins to slip into recently vacated brood cells in its immediate vicinity and to busy itself in them. It gnaws the cell walls with its mandibles and licks them with its proboscis. In doing so it is secreting something with which it polishes the cell wall until it looks smooth and shiny. This gets the cells ready again, either for the queen to lay more eggs in or to receive stores.

The young bees now feed from the cells of stores, especially on pollen (protein), yet they remain close to where they hatched. By making shivering movements with their wing musculature, which also has the effect of training the bees, they produce heat and thus help to maintain the brood nest temperature at a constant 35°C (95°F).

Bees between two and three days old begin to feed the older larvae of the bee brood with a mixture of honey, pollen and royal jelly. Over and over again they stick their heads in the brood cells for a few seconds before moving on to the next. In between they regularly collect food from the store cells.

After a further two or three days the bees devote themselves to cells with younger and younger brood, busying themselves in caring for them, feeding or inspecting them. At this age they can also join the bees that surround the queen for a time. The body of the mother of the hive is always being touched and licked. Bees are constantly offering her their proboscises so that she can take food from them.

In the period of life between the fifth and ninth or tenth day of development it is also possible to see lots of bees carrying out cell reconstruction or participating in capping them.

In the normal course of development, individual bees leave the hive for the first time around the ninth or tenth day. The young bees take their first flight to orientate themselves in the immediate surroundings of the hive. This is the orientation flight that was described in Chapter 2.

...to the outside

In the ensuing days we can observe the young bees taking the nectar from the incoming foragers and travelling to where the stores are being deposited. Hour after hour the fresh nectar, which at first is a watery liquid, is tended and thickened. To do this the bees take it up and expose it to the air of the hive. Over and over again it is sucked up and expelled. Each time a part of the water is evaporated in the warm, dry air of the hive. Then the honey is deposited in a storage cell to ripen further. Later the bees compress the pollen using their heads to stamp it down, and stick it together using saliva and nectar. As a result, it undergoes a lactic acid fermentation which breaks down its protein components and preserves it. Finally, it is covered by a layer of honey.

Subsequently, the bees may join a comb-building cluster and participate in exuding wax and moulding it into comb.

Around the eighteenth day of life, the bees may leave the building cluster and take part in cleaning and clearing out activities. They start 'guard duty' at the hive entrance, that is, they monitor incoming bees and try to drive away foreign insects. These can be strange bees from another colony, flies, moths or bumblebees. Frequently, in late summer, wasps try to get into the hive to steal the stores. It is often possible to see two or three bees jumping on an approaching wasp, tangling with it and trying to sting it. Another job at the hive entrance is ventilation. The bees hold on tight with their feet, many standing in formation one behind the other, and move their wings powerfully thus blowing a steady flow of air out of the hive. This airflow is warm and humid, especially on mild summer evenings after a particularly successful day's foraging, and it smells sweetly aromatic.

The bees stay a further two or three days round the hive entrance until they are about 20–21 days old, then they become foragers. When, at the height of summer, the development of the colony is still in progress they remain as flying bees for between ten and twenty days more until they die somewhere in the surroundings.

Metamorphosis of the glands

Scientists have investigated bees using lots of different methods and discovered that the course of development of these externally observable activities is accompanied by changes in physiological processes within the bees. They have various glandular systems within them, which gradually develop, reach maximum secretion, and then regress or change their function. For example, when a young bee has completed its task of feeding the larvae, its royal jelly glands diminish and its wax glands develop in order to meet construction requirements. By the time the period of wax work has ended, the venom gland has reached its maximum output and has filled the venom sac.

In the sequence of the processes described above, the following glands are active:

- Thoracic salivary gland (cleaning cells)
- Royal jelly glands in the head (feeding older brood, young brood and the queen)
- Wax glands (comb building)
- Scent glands (fanning, scenting)
- Venom gland (filling the venom or poison sac)

During further development the royal jelly glands, together with the thoracic salivary glands, supply the enzymes that enable the bees to refine the gathered nectar into honey. They remain active to a lesser extent in foragers.

It should be emphasised that our attempt here to describe

the development of the bee from hatchling to flying bee is statistically ideal, at least regarding its temporal components. In contrast to the development times for the individual brood stages from egg to hatched bee, which are fixed to within a few hours, further development is subject to the inner requirements of the overall colony organism, and therefore varies to a large extent. Certain developmental stages may be completely omitted if at the time there is little or no need for them. Wax glands for example do not reach full productivity in all bees when the space available for comb-building is filled and only capping or

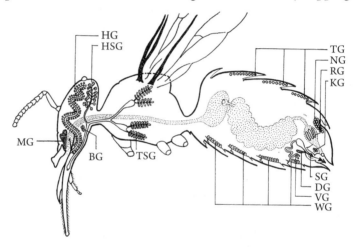

Cephalic or head glands:
HG = hypopharyngeal (royal jelly) gland
MG = mandibular gland
HSG = head salivary gland (head labial gland)
BG = buccal gland

Thoracic glands:
TSG = thoracic salivary gland (thoracic labial gland)

Abdominal glands:
DG = alkaline or Dufour's gland
NG = scent or Nasonov gland
VG = large, acid or poison or venom gland
RG = rectal gland or pads (rectal papillae)
TG = tergite glands (Renner Baumann glands)
KG = sting chamber or Koschevnikov's plate gland
SG = sting sheath gland
WG = wax gland

Figure 33. Schematic representation of the glandular system of a honey bee

remedial work needs to be done. Also, when vigorous foraging is in progress as a result of favourable conditions, the house bees that have not gone through all the stages become flying bees. Conversely, flying bees may reactivate their wax or food glands and resume the work of house bees if the development of the colony requires it.

Observation over a longer period shows that this aspect of the life of honey bees is only partially subject to an inner principle, and involves an active and reactive relationship with outer conditions and requirements.

Interpenetration

In considering the overall development of single bees amongst all the bees of the same age, another movement can be discerned. At first the bees stay in the hive close to where they hatched. Then they tend to move further and further away, gradually going through all the activities of looking after the brood. Then comes the orientation flight of the young bees, after which they are more active in the outer region of the inside of the hive: for example, managing the stores and depositing honey and pollen stops at the margins of the brood nest; and participating in the building on the margins of the existing comb. Cleaning, ventilation and guarding the hive entrance take the bees more and more towards the outside. Then, as flying bees, they repeatedly disappear into the infinitude of the surroundings, eventually staying there forever by physically passing away.

Thus, in the hive, we can experience two meeting and interpenetrating movements. From one direction comes the flow of substances that are constantly being brought in and there internalised. In the other direction a more selective passage through the hive is observable through the activity of the queen. As she is fed and tended she is constantly producing new eggs, which become spherically distributed in space as the area of the brood expands outwards through the hive. This metamorphosis of the queen's activity into the enormous metabolic output of the feeding bees and the developing brood, proceeds in the worker bees' expanding sphere of activity. Initially it is more inwardly orientated, later more outwardly developing, and ultimately culminates in the physical dispersal of the bees as they spread out into the surroundings of the hive.

7. Summer Impressions

The air is filled with the sound of buzzing.

To listen with closed eyes gives the impression of standing under a great bell that is being firmly rubbed. The whole apiary is humming. We might recall the experience of the swarming bees, except that this time the bulk of the bees are not flying around looking for something, but instead going directly to a place that is out of sight. Bees hurry from every hive entrance, take off and fly away quickly. Meanwhile, others descend as points of light out of the blue sky, briefly hovering in front of the hives before disappearing through one of the entrances. There is something overwhelming about this humming sound. On this hot, sunny day, the activity of several thousands produces a feeling of elation, like an inner rejoicing.

Beekeepers say of such a phenomenon, 'There is a flow on.' This is when everything that can fly rushes into the surroundings and returns again fully laden with nectar. This is likely to happen around midsummer when the lime trees are in flower or when the trees are producing honeydew; that is, when the leaf aphids in oak, sycamore or lime, or the *Lachnidae* in pine or spruce, secrete a sweet clear liquid which drips on the leaves or needles and forms a shiny sticky film. This is honeydew and many insects, including midges, hover-flies, wasps, ants and, of course, bees, come to lick it or suck it up.

About seven weeks have passed since the swarm under

observation was hived in its new box. The queen has twice laid up the comb that in the meantime has grown further. On some days she has probably laid more than 1,500 eggs in newly constructed cells or in cells vacated by hatching bees. On such days she lays at least one egg a minute and the total laid in 24 hours weighs more than the queen herself. In her large abdomen the ovaries comprise up to 300 strands, the ovarioles, in which new eggs are continually being formed and matured one behind the other. They then slide along the oviduct into the vagina. The secretory duct of the spermatheca has its outlet there. In the spermatheca is the sperm store which is filled with between 5 million and 7 million spermatozoa after the mating with drones on the nuptial flight. This store usually suffices for the whole life of the queen, which can last for between three and five years. How the sperm stay alive and undiminished in their fertility for so long is a question that has so far not been answered by biologists researching the matter.

When an egg passes the outlet of the spermatheca sperm are released, but only if it is being laid in a worker or queen cell. Drone cells are laid in without fertilisation of the eggs. This means that drones develop from unfertilised eggs and are thus fatherless. This process is called parthenogenesis. It is relatively rare in the natural world and thus the production of drones amongst bees is a biological curiosity.

It is possible to make a number of new observations if the hive is inspected around this time. The first thing to notice is that there is now comb in all the frames. Although comb construction in the frames at the sides is not complete, honey is already being stored in the outer combs together with freshly brought in nectar, and on the adjacent combs there is already some brood. All brood stages are well represented. Above the brood nest fresh nectar and pollen is being stored and, above this, the comb has been drawn out wide and contains a ring of honey for long term storage.

Bee dances

The house bees are busy everywhere with the activities already described, but some of them are noticeable for their special behaviour. They move round the comb in a pattern that always returns to the same point. They waggle their abdomens rapidly from side to side as they move in a short straight line then return to the starting point in a semicircle only to repeat the 'waggle' dance along the straight line, this time completing the other half of the circle on the return, and so on. There are quite a lot of bees showing this behaviour, especially in the lower part of the comb. Many are carrying pollen baskets. Some are walking vertically downwards on the comb, others upwards, many horizontally to the left or right, and some diagonally upwards or downwards. But all of them keep to their own direction and exhibit this waggling movement of the abdomen as they walk along the straight line. Some trace out the pattern of movements more vigorously and more frequently, others more deliberately and less frequently. The 'dancers' often repeat their figure of eight for several minutes without a break.

Quite a lot of bees on the comb are paying attention to them, touching them with their antennae, following them briefly and then going away again. This behaviour is reminiscent of that experienced on a hanging swarm until it eventually flies to a new site or a new hive, and we observed it with our own swarm before it was hived. Bee scientists call this behaviour in the hive 'neural communication'.

Karl von Frisch (1895–1981) studied these phenomena in a long series of experiments over several decades. He discovered that bee dances are connected with a particular foraging site found by a bee on one of her flights, and he observed that after she has returned to the hive more bees from the hive soon seek out the same site.

Frisch did experiments with dishes of syrup placed at various distances and in different directions from his observation hives.

He marked the returning bees with coloured identification dots so that he could observe their behaviour in detail on their return to the hive. He discovered that on their return to the hive the bees that had found a source of forage stimulated other bees to fly to the same source with the dance described, and he succeeded in decoding the dance pattern.

There are two distinct components to the dance. One is the direction of the straight line on the comb. The other is the frequency and intensity with which this distance is traversed during the waggle dance.

Because their eyes can detect polarised or orientated light, bees are able to sense in flight the direction of sunlight. They can do this even when they cannot actually see the sun itself, just as long as there is sufficient daylight.

As they walk on the comb in the darkness of the hive, the bees sense the direction of gravity. If the dancer moves vertically downwards in the gravitational field, the dance stimulates the attendant bees to fly away from the sun when outside, into the darkness so to speak. If the dancer moves upwards against gravity, it stimulates the bees to fly towards the sun. Upwards to the right and downwards to the left mean fly to the right in relation to the sun, either in the direction of the light or away from it. In this way every angle of flying direction in relation to the light source can be danced on the comb.

With dancers that sometimes dance for hours to stimulate bees to fly to a particular source of forage, the direction of the dance on the comb changes with the position of the sun in the sky, without the bees having to leave the hive. If a bee is dancing upwards in the morning, which directs flight towards the sun, at noon it dances to the left on the comb – that is, lightness/lighter to the right, heaviness/darker to the left – and directs flight at right angles to the sun's position with the sun on the right. In the evening it would dance downwards in the direction of gravity or heaviness, and thus direct flight away from the sun.

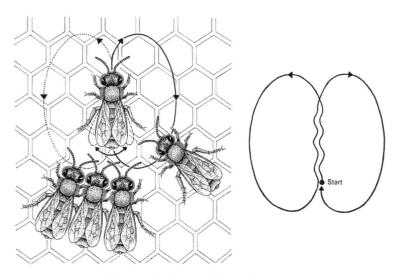

Figure 34. Pattern of the waggle dance on the comb

In watching the second visible component of the waggle dance, we noticed that different bees cover the straight course through the dance pattern at different frequencies. A bee with its foraging site closer to the beehive completes her passes through the dance more frequently than a bee whose forage site is further away. Thus, not only are the attendant bees stimulated to fly in a particular direction, they are also told how far to fly in that direction.

The dancing bees also carry with them the scent and taste of the forage source, and even the sugar content of the nectar and the yield of the source is conveyed by the vigour of the waggle.

Thus, the stimulus for the outbound flight is so exact that bees setting off for the first time do not have to hunt around for the forage site, they already know where it is.

With respect to the swarm suspended out in the open it should be noted that in this case the dances described stimulate the bees to visit the cavities that have been found, in order to assess their suitability as new accommodation for the swarm.

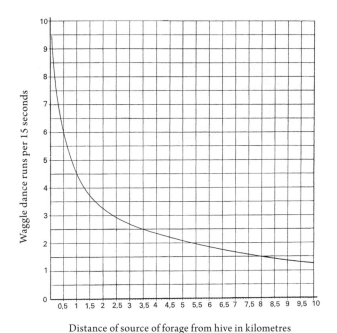

Distance of source of forage from hive in kilometres

Figure 35. Frequency of waggle dance cycle versus distance
to the source of forage

The swarm then flies away from its starting place when a large
proportion of its bees have danced the dance for a new site and
at the same time have got to know it.

To summarise once more the visible elements of the waggle
dance and what they mean:

DIRECTION OF MOVEMENT = DIRECTION OF FLIGHT

Down/with gravity/heaviness = Towards darkness/away from the light

Up/against gravity/lightness = Towards light/away from the darkness

> *FREQUENCY OF DANCE CYCLE = DISTANCE FROM HIVE*
>
> Higher frequency = Closer to the hive
>
> Lower frequency = Further from the hive

Here it may appear as if the terms 'closer' or 'further' are expressing something about the sensations of the bees in relation to the rest. Viewed from the centre of the hive, the immediate spheres may be sensed as being 'closer together' and the distant spheres as 'further apart'. This may also be expressed in the frequency of the waggle dance.

Drone developments

Apart from the phenomenon of dancing bees, further inspection of the swarm around this time reveals several combs of drone cells, as discussed in Chapter 5 (see page 47). The drones' eggs observed at that inspection have since hatched out of the cells, which have recently been laid in again. Others are still capped. The capping curves upwards above each cell like a dome so that the drone comb area looks noticeably thicker than the worker brood.

8. Queen Development

Further inspection around this time reveals two cone-like outgrowths suspended at the edge of the lower third of a comb. They are not quite as thick as a little finger and are about 3 cm (1 in) long. Their surfaces are covered with small depressions and look as if they have a hammered finish. In the bottom of one of them is a round opening into which a bee has slipped almost completely, only a part of her abdomen is still protruding. The other cone is capped with a small dome of wax. Now the opening of the first has been vacated by the bee and a light, gleaming white mass is visible in which a round larva is lying. It is comparable with a four- or five-day-old worker larva.

These are queen cells in which young queens are being reared. The colony is about to supercede its queen. The present queen, who is now three years old, is failing and, sensing this, the colony sets about replacing her.

A queen bee constantly secretes a pheromone – a hormone-like scented substance – through special glands that are situated predominantly in her head. This is received by the attendant bees and distributed throughout the hive via the uninterrupted social exchange of food that takes place between all the bees. If this subtle flow of queen substance decreases in strength, it can lead to profound changes in the further development of the colony, such as the aforementioned supercedure.

Figure 36. Queen cups (left) and a capped queen cell (right)

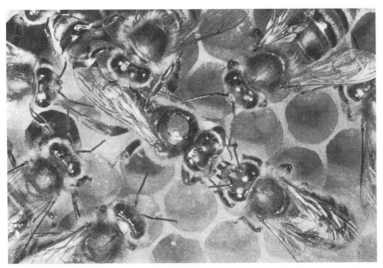

Figure 37. While she is being tended the queen is constantly being sensed by the workers

If the queen is lost for some reason, either through a clumsy manoeuvre by the beekeeper or if she is removed, the bees realise this within half an hour as a result of the interruption of the pheromone flow. They sense that they are queenless. They run all over the place looking for her and soon begin to roar loudly. Beekeepers refer to this as 'moaning'.

Figure 38. Social feeding between two worker bees

If such a colony has access to open worker brood, some of the cells are drawn out into 'emergency queen cells' which are similar to normal queen cells. The younger the larvae in these cells the better the emergency queen that is produced from them by the bees. The older the larvae – that is, beyond three days – the smaller the number of ovarioles that will reach full development and the lower the laying capacity of the queen that will result. This is connected to nutrition. The older worker bees are fed with a mixture of honey, pollen and a small amount of royal jelly. This ultimately turns or moulds them into worker bees through the development of the appropriate glands and gathering organs, and through suppression of development of the specific female reproductive organs.

By contrast, a larva developing in a normal queen cell is supplied with large amounts of royal jelly, the white substance seen covering the whole cell 'floor' (ceiling in the hanging cell) and in which the larva lies (hangs). The nurse bees feed the larva in a queen cell so generously with royal jelly that she floats in a surfeit of food, and after the hatching of the 'princess' a great deal remains dried up in the cell. During the five days of her development before cell capping, the queen larva is fed about 1,600 times. Beekeepers only find these larvae floating in royal jelly in their first three days, but the worker and drone broods are given only as much food as they can consume. Thus, a worker larva is fed only 143 times in the six days of her development, that is, approximately hourly.

Figure 39. Emergency queen cells are drawn over cells with open worker brood.

Besides the feeding there is another factor to take into consideration which distinguishes the development of the worker from that of the queen. Worker bees grow and develop over 21 days in hexagonal cells in the body of the comb. Queen cells are rounded from the outset. A queen cup, usually situated on the edge of the comb, is at first almost spherical with an opening at the bottom. Then it is expanded. If the queen lays in it and the larva develops then it is gradually drawn out into a cylinder and ultimately capped with a hemispherical dome at the bottom. With all these differences it is hardly surprising that the period of development of a young queen has its own rhythm.

After three days of the egg stage and five days of the larval stage the cell is usually capped, and after eight days of metamorphosis the queen emerges: that is, sixteen days after the egg was laid.

Queen (open queen cell / capped queen cell)

Age in days	Beekeeping term	Moults (ecdyses)
1		
2	Egg	
3		hatching
4	Round larva (or grub)	1st moult
5		2nd moult
6		3rd moult
7	Stretched larva	4th moult (capping)
8		
9		
10	Pre-pupa	
11		5th moult
12	Pupa (nymph)	
13		
14		
15		6th moult (hatching)
16	Queen	

Worker

Age in days	Beekeeping term	Moults (ecdyses)
1		
2	Egg	
3		hatching
4		1st moult
5	Round larva (or grub)	2nd moult
6		3rd moult
7		4th moult
8	Stretched larva	(capping)
9		
10		
11	Pre-pupa	
12		
13		
14		5th moult
15		
16	Pupa (nymph)	
17		
18		
19		
20		
21	Worker	6th moult (hatching)

Drone (open brood / capped brood)

Age in days	Beekeeping term	Moults (ecdyses)
1		
2	Egg	
3		hatching
4	Round larva (or grub)	1st moult
5		2nd moult
6		3rd moult
7		4th moult
8		
9		(capping)
10	Stretched larva	
11		
12		
13	Pre-pupa	
14		
15		5th moult
16		
17		
18		
19	Pupa (nymph)	
20		
21		
22		
23		6th moult
24	Drone	(hatching)

Table 2. Overview of the brood developmental stages of queen, worker and drone

As a rule, most of the queen cells are drawn in a colony that is in swarming mood. Frequently, between one and two dozen larvae are developed into young queens. From a biological point of view, in strong, healthy and vigorous colonies, swarming is the method by which the colony reproduces, rejuvenates and spreads itself further. A prime swarm with a queen that has been active in the colony until that point will normally emerge on the day that the first swarm cell (a queen cell produced with the bees in swarming mood) is capped, unless it is delayed or prevented by unfavourable outer or inner conditions such as bad weather, forage limitation or a queen that cannot fly. In other words, it emerges on the ninth day after the queen has laid an egg in the first queen cell. A week later, with the subsequent hatching of the virgin queens, several secondary or after-swarms can occur, often referred to as 'casts.' These may continue until the inner conditions of the colony bring it to an end through the killing of the unhatched young queens. This may be carried out by the last queen to hatch, who bites into the remaining cells, or by worker bees. The remainder of the colony has at its disposal all the comb with the stores it contains. The number of bees in it is replenished in the ensuing days by the hatching brood. About five days after hatching the young queen flies once, or several times, to the drone congregation area to be mated. But she only starts to lay when the last brood from the old queen has hatched.

III. WINTER IS COMING

9. Maturity

The next inspection of our swarm, which has now become a strong colony, takes place about two and a half weeks later. Midsummer has passed. The lime flowers are fading and the woodland forage is drying up. Despite the good weather, the foraging traffic is rather indifferent. 'Beards' of bees are hanging in front of the entrances or from the alighting boards of some hives. These are bees that have nothing to do in the hive and who would produce too much heat in there, and yet cannot find any forage worth bringing in. Some bees are flying around, noticeably restless, as if looking for something around the hive, and hover up and down by the bees in front of the hive entrance. Sometimes they try to get into a hive but are immediately driven away by the bees at the entrance. Wasps are seen behaving in this way too.

The bees doing this are called 'robbers' by beekeepers. They have become redundant in their own colonies because of the shortage of forage, and now they are trying to obtain from other colonies what they cannot find in the environs. Unfortunately, if a colony is not strong enough to defend itself it will be robbed clean within a few days and collapse. In order not to provoke robbing, inspections at such times are brief. Anything needed for knowing what measures are still to be taken for late summer management must be completed in two or three minutes.

The outer frames are now fully built out with comb and these are filled from top to bottom with honey. The cells are already largely capped with wax. This tells the beekeeper that the honey is ripe and can thus be stored for a long time. The two full outer combs on the right and left hold about 2.5 kg (5 ½ lbs) honey. This could now be extracted but it would interrupt the colony's development and so, for the time being, it is left alone. Also, by leaving it to the bees the beekeeper is spared having to supplement their winter stores by feeding them sugar, or at the very least it reduces how much they have to be fed.

The comb with the queen cells is carefully removed. It is immediately clear that one of the two queens has probably hatched as the bottom of the cell is round and open and the cell is otherwise intact. The other is still closed at the bottom but it has a big hole in the side through which it can be seen that the cell is empty. It has probably been chewed out.

A queen is walking about on the adjacent comb. The new queen looks very large by comparison with the workers around her and is obviously already laying. She has just stuck her head in a cell and is now turning round and introducing her abdomen into it. She dips her body backwards into the cell up to the middle segment, steadying herself sideways with her reddish-brown glistening legs. A moment later she rises and wanders off in search of another cell to inspect. As she does this, she is constantly surrounded by a circle of bees who look after her and feed her. A longish, white egg can be seen at the bottom of the cell she has just left.

The beekeeper carefully replaces the comb in the hive and inspects another closer to the outside to get an overview of the colony's stores This comb has a broad ring of honey surrounding the brood nest. A bee with a coloured tag on its back is visible. This is the old queen. The inspection completed, the beekeeper places the comb back into the box and carefully closes up the hive. It has already become noisy around the hive as robbers try to gain entry from above.

It is rare to see two queens in a colony. Normally a colony will support only one queen, as rival queens would try to kill each other. The old queen will only carry on her activities for a while beside the young one in colonies that supercede the queen.

The drones depart

Visitors to an apiary after midsummer may observe something else. It is early morning. The dew is on the grass and the summer sun is climbing over the horizon above the mist. The birds are still singing but they are more muted than a couple of weeks before. Small puddles of condensation lie on the alighting boards of the hives as evidence of the processing of the nectar during the preceding night and the vapour from the air being blown out of the hive.

Figure 40. Drone being dismembered by a wasp

A bee drags a round, longish, white object out of the hive and lets it drop from the alighting board. A lot of these white bodies, each about as thick as a pencil and at least 1 cm (½ in) long, can be seen on the ground in front of the hives. Closer inspection shows that they have parts which are vaguely suggestive of bees: a head with eyes, antennae, mouth parts, bodies with folded-in leg segments, abdomen with light rings ending bluntly. But they are colourless, somewhat translucently white, occasionally light brownish. Some of them have eyes tinged with violet. These are drone brood that have been cleared out of the combs by the bees. A few round wax cappings scattered around suggest that the bees have uncapped the brood, pulled out the motionless pupae and carried them out of the hive.

Around noon on this day the drone traffic is noticeably less than it was a fortnight earlier despite the sunshine and hot weather. Some of the colonies have no more drones flying whereas others have only one now and again. They are driven away from the hive entrance by the workers if they try to get in. The guard bees push them away with their bodies, and tear the wings and legs of these ponderously crawling, big-eyed droners with their mandibles. Hardly a single drone manages to gain entry into any hive whose bees are showing this behaviour. Many fly off to another hive and try to slip in there. Others remain sitting below the alighting board where in the cool of the ensuing night they become moribund and drop to the ground. Tits, sparrows and other birds help themselves to the pupae and fly off with them. Wasps fly around the hives and dismember the defenceless drones with their mandibles. The head, legs, wings and abdomen are separated from the thorax which is then carried away.

After a few days there are no more drones flying from this apiary into its surroundings. This as an indication that the colonies there are 'queen right', in other words, at this time of year they can sense that they are inherently harmonious, each with a vigorous queen at their centre. If a colony still has some drones in late summer, or is still tending drone brood, it can be assumed that it intends to supercede because, for whatever reason, it is dissatisfied with its queen. A colony in Central Europe starts raising drones at about the beginning of spring. Thus, the first drones hatch around Easter and are able to fly ready to mate at the end of April or beginning of May. In early localities, hives may throw prime swarms at this time. By the end of July or the beginning of August most of the drones have disappeared again, and in Central Europe swarms are not normally expected after that time.

Wintering

A few weeks later the heat and drought of summer is well past its peak; the grain-fields have been harvested and are being prepared for the next crop; the clusters of berries on the rowans are bright red; the first apples are already rotting under the trees and the wasps are flying close to the ground looking for food. They will find the smallest holes to get into the hives. The beekeeper has reduced the size of some of the hive entrances to make it easier for the bees to defend themselves against any wasps that seem intent on robbing them. Yet for a while in the slightly misty, humid coolness of daybreak they can penetrate the hive unchallenged. However, as they quickly come out again and fly onwards to hunt elsewhere, their success is somewhat limited. Nevertheless, any dead bees and other debris thrown out of the hive is torn apart and carried away by ants and wasps.

Even on hot days the bee traffic in the apiary is rather moderate and subdued. The bee beards, often seen at hive entrances at the end of July, have now disappeared. A look in the swarm colony that we have been following shows that the number of combs with brood has decreased. It is still to be found on about five to six combs. The brood area has decreased too and at least three quarters of it is capped. From this it can be concluded that the queen has been laying fewer and fewer eggs, and that consequently there will be fewer replacement bees than before to develop from these.

The outer combs are visited now by only a few bees and are largely filled with capped honey. Even over the brood areas there are wide rings of honey, mostly capped. The brood is surrounded by a ring of cells filled with pollen and some of these are covered with glistening honey.

The beekeeper makes another survey of the colony's stores and finds them adequate for over-wintering. This colony does not need supplementary feeding with syrup. This is not always to be taken for granted. Without the support of artificial feeding many swarms do not manage with their own resources, either to build sufficient comb or to produce and care for enough brood to create a colony strong enough to over-winter, or even to bring in sufficient stores to continue. The colony has to succeed in bringing about a balanced relationship of comb, brood and stores. For one thing, this is dependent on outer conditions of the apiary site, such as microclimate and forage development, as well as on weather outcome. For another, it is influenced by a number of inherited traits as well as the individual strength that the colony can summon up out of the act of swarming.

Figure 41. Bee sting with droplet of venom

Figure 42. Comb with winter bees below capped honey stores

In order to follow the further development of the swarmed colony, the next inspection may take place around Michaelmas. By then the leaves are beginning to change colour and the October sun creates a multitude of warm, earthy tones in the surroundings. The days still quite mild, but the nights are noticeably colder. The dew stays on the plants late into the morning; spider webs shimmer in the morning sunshine, which slowly disperses the light ground mist lying in the hollows. Even in the warm hours around noon, there is only weak foraging traffic. Pollen gatherers are in evidence on Michaelmas daisies and goldenrod; otherwise there is merely a brief orientation flight of young bees.

In the colony, the outer store combs are almost abandoned. A thick coat of bees is now gathered on and under the combs where before we saw the brood nest. The bee cluster is suspended between the seams of comb and hangs down almost to the hive floor. A carefully removed comb reveals closely packed bees interleaved in two layers like roof tiles, their abdomens pointing downwards and outwards, their heads and thoraxes pointing inwards and upwards. This coat of bees moves only slightly and the colony flares up a bit on being disturbed. A few bees point their abdomens in the air and the outstretched stings can be clearly seen with tiny droplets as clear as water on the ends of them. This is a small amount of bee venom.

There is no more brood, or at most small patches of capped brood cells on one or two combs at the centre of the cluster of bees. The remaining young bees will soon hatch from these. This shows that the queen laid the last cells around the equinox.

A colony's winter cluster forms over the vacated area of cells of the previous brood by hanging in the gaps between the combs close to the hive entrance. The uppermost bees of the cluster occupy the lower margin of the ring of honey of the comb they are on. A colony in this condition is ready for winter.

Unless there are special circumstances, the beekeeper does not need to open the hives again before the following Easter. The entrances are restricted or fitted with protective grilles (mouse guards) to stop mice getting in. A sampling board is inserted into the floor of the hive under the bee cluster. This is made up of a water-resistant sheet beneath a fine mesh, which is designed to be removed without opening the hive. By examining what falls from the over-wintering cluster onto this board the status and condition of the colony can be determined without having to disturb the hive.

10. Winter Inspections

The mood at the apiary, even during daytime, is damp, silent and gloomy. The leafless branches of the trees and shrubs look wet and slippery. When it is not raining, landscape and sky are frequently obscured by fog. The sun rarely shines directly onto the ground. Birds only occasionally fly past; at most we notice a brief flitting amongst the branches and a soft piping or short twitter. Occasionally, a tit comes to the apiary and quickly picks up a bee that is lying torpid on the alighting board or on the ground.

It has to be very mild with the air temperature over 12°C (53°F) to tempt at least a few bees out of the hive. These fly around as the young bees do in their orientation flight. Some of them visibly void their recta as a spray, or deposit a brown blob on the front of the hive. In a healthy colony, bees never defecate inside the hive. Before the anus is a greatly expandable rectum in which indigestible food residue and excretion products gather until a bee can make a voiding, or cleansing, flight to deposit the faeces in the surroundings. The volume of the rectum is normally sufficient even for the long period of winter when the bees cannot fly out because of low temperatures, which can often last for three months or more.

During this period a beekeeper makes mostly external checks to ensure that everything is in order. The hive is not opened. The sampling board is inspected and cleaned. There are a few bees lying on the mesh, mostly hunched up with their legs drawn in. These are the bees left from the summer that have reached the end of their life-span during the preceding days. The mesh is lifted out and the bees brushed onto the grass.

Reading the debris

There are some strips on the sampling board that create the appearance of a circle: the outer lines are thinner and shorter, the inner ones broader and larger. They are made up mostly of fine brown crumbs, and when rubbed between the fingers they feel soft. They are crumbs of wax. This comes about when the bees in the upper half of the winter cluster gnaw away the cappings of the honey cells to get at their stores. The crumbs of wax fall down between the combs. Because the bees form a broad sphere between the combs, this gives rise to the circular pattern formed by the strips on the sampling board.

There are other bits in this debris and the experienced beekeeper can draw several conclusions from them regarding the condition of the colony without having to open the hive.

Dry cappings without significant amounts of other components is a good sign. Black crumbs and cocoon indicate presence of wax moth larvae living in the comb. Chewed bees indicate that a shrew has visited the hive (the musculature of a bee thorax is certainly a tasty meal for it). Mouse faeces and nest material such as dried grass and leaves, indicate that a field mouse has nested in the hive and has helped itself to bees and honey. Moist debris indicates that the colony is still raising brood. This increases metabolism and thus creates more water vapour. In that case, there are often pieces of dead bee brood in the debris that the bees have pulled out of the cells.

In Central European latitudes, beekeepers like to have brood-free colonies by November. Broods need a constant temperature of 35°C (95°F), which has to be maintained by the bees against very low external temperatures. The brood also has to be fed.

As a result, the bees do not cluster as tightly or as motionlessly. This increases the consumption of stores and has an unfavourable effect on the climate inside the hive because of the associated increase in humidity and the condensation of the water vapour. The bees maintain a temperature of 20–30°C (68°F–86°F) in the centre of a broodless winter cluster. This slowly warms the sphere to its outside, so that the outer bees do not become too cold. The bees slowly change position with one another in the cluster from inside to outside and from outside to inside.

Losing a queen

Sometimes fresh wax scales can be seen in the debris and occasionally eggs that the queen perhaps wanted to lay. If round drone cappings (cappings of drone brood) are found in winter, it can be assumed that the colony has drone brood. This means that it no longer has a fertile queen. This can occur if she was not mated, or mated insufficiently, on the nuptial flight, or if this flight was prevented by a long period of cold weather. Possibly she has exhausted her sperm supply as a result of old age or the eggs are not being fertilised because of some internal abnormality. It could also be caused by the loss of the queen. Perhaps she has died or was accidentally injured at the last inspection. In such an emergency situation the colony raises one or more pseudo-queens, or drone-laying workers. Workers are fed with royal jelly, which causes them to develop ovarioles of limited capacity, and these start to produce eggs. But these eggs are not fertilised and thus only produce drones.

A colony that is just producing drone brood is doomed. It no longer has the ability to maintain the flow of life on its own and will most likely collapse by the following spring at the latest. Remedial action can be taken by uniting it with another colony so that the remaining resources of the queenless one can be salvaged. But even this procedure is not always successful.

11. Varroa Destructor

Today, beekeepers look for another symptom in the hive floor debris of their colonies. This appears as tiny, dark brown oval platelets 1–2 mm in diameter. Under magnification they show antennae, mouth parts and four pairs of legs. Forty years ago this phenomenon was unknown to beekeeping in Central Europe, but today there is hardly a colony that does not have these platelets, and they can be found in hives over most of Europe. About fifteen to twenty such platelets are present in the debris of our colony.

Figure 43. Varroa mite (greatly enlarged)

These mites, or arthropods, are called 'varroa mites.' *Varroa destructor* were discovered and characterised on the East-Asian honeybee species *Apis cerana* at the beginning of the twentieth century. The mites live in bee colonies. The female mites slip under the tergites, the rings of a bee's abdominal segments, and suck out the haemolymph, the bee blood, and the fatty protein body of the worker bees. After a mite has matured for several days, she leaves the bee and looks for a brood cell which is at two days before capping (seventh to eigthth day of brood development). She slips under the round larva into the liquid food and there enters a kind of paralysis. She ends this inactive state only when the food has been completely consumed and the larva in the capped cell has become a stretched larva. She climbs onto the larva, punctures it in the side, and feeds on its haemolymph for about two days. From about the fourth day after capping, and at intervals of around thirty hours, a female varroa deposits protonymphs on the cell wall. This is a developmental stage between egg and fully formed mite. These moult for the first time after a further thirty hours and move to the bee pupa in order to feed on its haemolymph at the same puncture. The first protonymph is usually a male varroa, which then mates with subsequent fertile female mites (his sisters) before the bee hatches, and then dies. The deposited protonymphs take about six days to develop into viable varroa.

About three varroa females can grow on a hatching worker bee, although usually only one of these mites is fertile. In drone brood, which of course requires 24 days to develop (fourteen days as capped brood), there may even be as many as five or six varroa of which three or four may be fertile. After their proper period of maturing on living bees, they again move into the brood to reproduce. During her lifespan the varroa female can infest several brood cells (between six and eight are possible). Unmated varroa females enter brood cells to breed, but the protonymphs develop into males only. The female is then fertilised by the first sexually mature male. After that she can reproduce normally.

Figure 44. Varroa on a bee pupa

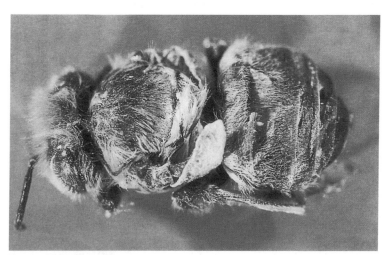

Figure 45. Deformed bee

Bees hatching from infested cells are usually weakened. With higher varroa infestation, two or more varroa may enter a brood cell. The hatching bees then show deformities of their wings or legs, or the brood even dies in the cell.

The reproduction of the varroa mite that parasitises *Apis mellifera* – the honeybee native to Central Europe and now distributed worldwide – generally takes a progressive course. This means that the number of mites in a colony increases exponentially, doubling in size every three or four weeks. From a single infestation of a few mites, over the course of six to eight months they become several thousand. Individual bees and brood of the infested colony are so weakened by this that other infections by bacteria or viruses lead to disease. Such colonies usually collapse in a short time in late summer or early autumn when the colony is preparing for winter. Apart from a few bees and dead brood the beekeeper suddenly finds empty hives.

The collapse is actually a disintegration of the hive which loses its inherent cohesion and integrity that forges the thousands of single bees into a single entity. The collapse is an important opportunity for the varroa mites to spread. The mites are not mobile and cannot move away from the colony on their own. However, the large numbers of departing bees transport them to other colonies which they infest, mixing and developing with mites there.

The problem is especially serious because European honey bees have not yet been able to adapt their behaviour to limit the development of these mites. It is quite probable that honey bees would have been almost wiped out here had it not been for the fact that beekeepers, scientists and the chemical industry rushed to find ways of controlling mites in the colonies. In the last twenty years the number of mites has usually been reduced using organic acids as these leave no dubious residues and the mites are not expected to develop a resistance. Some beekeepers are using treatments to disrupt the development cycles of the mites.

A new population of honeybees seems to be slowly developing that has some kind of regulatory ability. Some factors in this are known, but a genetic disposition to carry a resistance to varroa into the next generation has not yet been observed.

It is now absolutely necessary to carry out varroa control measures on each hive at least once a year.

12. Bee Christmas

Bees do not forage in winter time because they will not find anything in the frozen countryside. They pack themselves together and now and then take a tiny drop of their stores. The warmth of the summer sun is stored in the honey and is set free in the body of the bees, and so the bees are fed and warmed at the same time whilst they guard their queen in their midst. A winter cluster moves very slowly between the combs towards the stores thus gradually getting further and further away from the hive entrance.

Beekeepers often put an ear to their beehives in winter to listen. Through the wood a soft humming, whispering, scratching and scraping can be heard. If the wood is knocked with the knuckles there is a roar from the colony, which soon falls quiet again (only a queenless colony moans its complaining reply with a longer rising and falling sound). The colder it is outside the louder the humming. It is also louder when the colony has brood again.

In former times, beekeepers would tell the bees the Christmas story. Today this custom of communication between people and animals has largely been forgotten, and even those who know about it and sense its deep significance often have difficulty in going about it in the right manner. However, this encounter can give the animal world a feeling of confidence and security, which is what being a domestic animal is all about: being commended into the care of human beings. *And so, at the darkest time of the year, the bee organism dreams of Christmas, and through this dream renews its power to safeguard the secret of maintaining the light, the warmth of the summer sun, in honey.*

Now and then we notice a bee leaving the entrance and taking off, even in cold weather or when there is snow on the ground. She flies for a short distance and a few metres high only to drop to the ground. She moves for a while and then quickly becomes torpid. If she falls on snow, we notice that she sinks into it a little, sometimes one or two centimetres deep.

It is claimed that this phenomenon occurs during winter when a bee realises that she is going to die and departs the hive so as not to burden the colony with her corpse. But we might also imagine that *with her life the bee blows a greeting of summer sun from the bee colony into the wintry world: Spring is on its way!* [3]

IV. BUILD-UP

13. Awakening

On 2nd February, exactly six weeks after the winter solstice, the church celebrates Candlemas, the Feast of the Purification of the Virgin Mary. This is a significant date in farming and in phenomenological calendars for a number of reasons. At this time the sun is shining for an hour longer than it was at Christmas, and it no longer gets dark so early in the afternoons. The songbirds have become more active and we can hear them singing again. Bud growth is visible on trees and shrubs and the stumps of trees, and branches that have been cut around this time often begin to 'bleed' profusely. Plant sap is exuded and is frozen by the frost into little fountains. In Germany, the first mild days come around this time when the sun at midday heats the air to over 12°C (54°F). Hazel catkins start to flower and their yellow pollen is blown away by the wind.

After the brood-free period in the bee colony, the queen begins to lay the first eggs in the cells. She then starts to lay every freshly vacated storage cell increasing the area in an expanding circle so that the brood nest grows in the wake of the receding stores.

Cleansing flight

Beekeepers visiting their apiaries remove the mouse guards and take out the sampling boards to clean and inspect them. The first bees will probably have come out by now and soon many more will follow. There is an increasingly lively humming around the whole apiary, like a vigorous orientation flight of young bees. The bees fly in loops and lemniscatory forms of increasing size. At the same time, they empty their recta, some in flight, some after alighting somewhere, especially on bright, gleaming surfaces. In residential areas, beekeepers have to ask neighbours who have hung their washing out to dry to take it in again, otherwise they would wonder how their washing suddenly acquired a new pattern. The bees even land on observers and deposit blobs that are ochre to brown in colour.

Figure 46. Bee foraging in spring

Bees drag winter debris and the corpses of other bees from the hive and fly off to drop them in the surroundings. Meanwhile, the first forager bees have found their way to the hanging catkins on the hazel bushes and are gathering the bright yellow pollen into their baskets.

But as soon as it cools down, for example when a cloud covers the sun, the flight traffic stops and the bees hurriedly disappear back into the hive. Otherwise the cleansing flight continues for at least another hour.

In the days that follow, bees can frequently be seen hurrying out of hive entrances. When it is mild, they collect flower pollen and the first nectar from the early spring flowers, but even in bad weather they can be seen gathering water nearby. This is an indication of the increasing amount of brood being reared in the colonies. An inspection of hives at this time would show, in the centre of the bee cluster, a brood area that might extend to two or three frames.

The time of the cleansing flight is a joyful one for beekeepers. They see that their colonies have survived the winter so far and are becoming active in the open again.

14. Surviving Spring

After the cleansing flight, peace descends upon the apiary once more. The afternoons are still cold this early in the year and the nights often bring frost, especially when the days are sunny and clear. The ground remains frozen where snow is lying and there may be more snow or perhaps even hail tomorrow.

Although the light is increasing day by day, we cannot be sure of stable weather conditions, even after the spring equinox. After a brief warm period comes wind, rain, cold or snow; the lighting conditions change constantly: it is typical April weather.

The warmth of the sun can entice bees out into the open, but a cloud with a gust of wind and a shower of snow can throw them to the ground where they become torpid. If the sun comes out and warms up the fallen bees they might stir again, but only a few of them become airborne once more and make it back to their hives.

Beekeepers watch these events with concern. Spring is a critical time that determines which colonies are healthy and strong enough to build up their resources through the Easter period and survive into summer.

Dangers

Inspecting the colonies at the beginning of March presents us with the following overall picture. The outside combs to the left and right are still full of capped stores, whereas the ring of remaining food nearer the middle has become smaller. There are patches of brood on two frames, perhaps a third in rare instances. If the weather in the preceding days has been mild and it was possible to bring in hazel and alder pollen, there will be a big patch of open brood. But if it has been cold, even frosty, there will be less open brood. The queen in this case has restricted her laying activity, and if the weather stays cold for a long period, meaning the bees cannot go foraging, the queen will stop laying altogether.

When it becomes so warm outside that the hive itself is warmed, the bees begin to gnaw into the combs of stores at the outer margins of the hive. They collect the food and carry it to the inner region close to the sphere of brood. Depending on the consistency of the honey, more or less water is required for the bees to dissolve it and suck it up. Once the capping is removed the honey absorbs water from the cooler and more humid air around the margins of the hive and becomes more pliable. However, if the honey stores have crystallised out to a greater degree it is more difficult for the bees to take them in again. This can depend on the forage brought in during the preceding year. Honey from early foraging (fruit tree flowers, dandelion and especially oilseed rape) sometimes crystallises very coarsely and becomes hard. But several summer nectar sources (sunflowers, mustard, *Phacelia*), especially a predominance of a single one, or certain sources from leaves (honeydew containing melezitose), can also become so hard in the combs that it is difficult, or even impossible, for the bees to liquefy the honey sufficiently in order to take it in as food. A colony affected in this way faces starvation because it cannot mobilise its stores. As a result, it develops slowly.

There are other situations that are critical for the colony. When the total supply of stores is too small the bees will generally starve. The beekeeper can help them only to a limited extent by supplementary feeding. In either case the colony will come out of the winter weaker.

If the dome of stores over the bee colony is exhausted and the weather is extremely cold, the bees cannot move between one comb seam and another to get at the stores nearer the sides of the hive. In such situations they can starve even though they have full frames of stores nearby. The situation is especially critical when brood is already present and the increased scattering of the cluster among the combs draws it away from the brood. This splits the winter cluster apart. One part of the bee colony stays with the brood it is looking after and starves there, whereas the other – often significantly weakened – carries on feeding on the stores.

A beekeeper needs to find the cause of this development. Usually it is discovered that even before over-wintering there was not the right relationship between colony strength, comb development, hive volume and stores.

15. From Winter Bees to Summer Bees

A bee colony starts the winter with an average of 12,000 bees (between 8,000 and 15,000). These bees hatch from cells between August and October and form the winter colony. They survive the whole six months of winter until Easter. The last of the winter bees generally reach the end of their life-span at the beginning of May. This means that under certain circumstances winter bees can be eight months old whereas summer bees normally live for six weeks at most: 21 days as house bees and a further ten to twenty days as foragers.

If we study the development of individual summer and winter bees some marked differences can be noticed.

Summer bees, first as house bees, perform the full range of tasks inside the colony such as heating, feeding, nursing, exuding wax, building, and guarding the entrance. Then, after a further period as foragers, they die somewhere in the surroundings.

Winter bees hatch into a colony situation which can to some extent be described as a finished job. Foraging activity is limited as there is less nectar available and the supply is becoming exhausted; then too there are the shorter daylight hours and the cooling temperatures. Within the hive the comb has already been built, the stores have been tended and sealed up, and the brood needs feeding less and less. Thus, the bees hatching into these circumstances do not have much to do externally.

They feed themselves with bee bread (pollen stored in cells mixed with other substances, see page 138 for a more detailed description of bee bread), and perform, to a limited extent, the tasks in the hive that the house bees would carry out in summer. Winter bees form fat bodies, with pads of stored protein and fat under their abdominal wall. Research has also shown that there are changed hormonal conditions in the colonies at this time.

Only with the brood rearing that starts at the end of winter (February to March) do winter bees begin a development comparable with summer bees. Stores in their fat bodies are broken down, nursing brood increases and building activities resume. The more that new bees hatch out and go through their development as house bees, the more the winter bees become foragers. After a time, they stay in the surroundings.

Population estimates show that a summer colony contains between 35,000 and 45,000 bees. In rare cases there can be more than 50,000. Of those, at least one third are foragers, the rest are house bees. From towards the end of July (the end of St John's Tide) the number of bees decreases and the brood volume is greatly reduced, halving approximately every 4 weeks until the last brood emerges in October (under UK conditions, small amounts of brood are not infrequently present in the winter months). The summer bees gradually disappear leaving behind the winter cluster. A healthy colony loses only a few bees until the increasing daylight stimulates the production of the first brood. During this period, only a small portion of the stores is used, less than 1 kg (2 lbs) per month. As soon as brood is being reared, fed and kept warm, the consumption of the stores increases to between four and six times what it was during brood-free conditions.

More water is used in colonies with brood, especially for dissolving the stores, and there is increasing demand for fresh pollen. The early flowering trees are primarily hazel, alder and willow. The increasing foraging traffic combined with sudden changes in weather often leads to high losses of bees. The colonies fly themselves bare. If fewer young bees hatch in a colony than are lost, the colony becomes weaker. This has harmful repercussions for the situation inside the hive. If the brood is less well supplied the queen reduces her laying and fewer young bees are produced in the coming weeks as a result.

How a beekeeper can help

With increasing development, the size of the brood nest doubles roughly every four weeks from mid-March to mid-May; the critical stage of loss of the winter bees is well matched by the hatching of young bees, and the colonies become very strong.

As the frames of comb can be moved at will, beekeepers have a number of options to help the colonies in their development. In weaker colonies, the volume of the space can be reduced by removing empty unoccupied combs and separating the bees from the vacated space by inserting a movable dummy board, or partition. The smaller space can be better managed in terms of its warmth or climate by a weaker colony which, if it is healthy, is thus enabled to develop more quickly. Alternatively, if a colony is strong, the beekeeper may increase the volume of available space by giving it empty frames to build comb in, or supplement the existing frames of comb with additional ones. The beekeeper may also insert foundation frames. These are thin sheets of beeswax embossed with the proper hexagonal pattern (the same as those used to roll beeswax candles). These sheets are fixed vertically in the frames and the bees draw out the comb on them. The foundation sheet forms the midrib of the comb.

Depending on the hive system in use there is also the option to increase the size of the hive above or below the brood nest. This can be done by adding empty frames, or frames of foundation or comb, in another box at the top or bottom. These boxes are called supers. They are open at the top and the bottom, with rails to hold frames, and they are placed on top of the brood box under the roof. A strong colony then occupies this empty space, filling it with fresh comb and using it according to its capabilities and resources.

The beekeeper tries to get a picture of the colonies during each inspection, observing the flight traffic, the foraging, whether the incoming bees are also carrying pollen, the orientation flight of the young bees, and the behaviour at the hive entrance. Inspections in March are still infrequent, but in mild weather distinct differences can be observed between the colonies. The first drone flight is seen again around Easter before the cherry trees are in flower.

16. Preparing for the Honey Harvest

When the fruit trees begin to blossom, the colonies are checked to see which ones will be mature enough for storing honey for harvest. If the brood space is well filled with bees and the brood nest has grown large and can be supported satisfactorily, then it is high time to give the colony space to store honey. This is done by placing a super with drawn comb (see previous chapter) on top of the brood box, usually over a queen excluder. This is a sheet or grid of slots through which the queen cannot pass because of her larger body size. This prevents her from entering the super and producing brood there.

When the weather is favourable and lots of bees are visiting the early forage (flowers of fruit trees, dandelion, rape, sycamore etc.) this additional empty comb soon fills with honey. Sometime after Ascension Day or at Whitsun (late May to early June) the beekeeper may harvest the first full honeycombs.

First the combs must be uncapped. The wax cappings covering the cells containing ripe honey must be carefully removed with an uncapping fork or a knife. The honey is then spun out in an extractor. This involves placing the combs upright in a basket which then rotates at speed. The principle is similar to that of the spin-drier in a washing machine. Under the influence of the centrifugal force the honey flows out of the cells and flies into the extraction vessel where it is collected. These spun-out combs can then be given back to the bees.

Figure 47. Full comb of honey – all cells are capped

If there is a good amount of summer forage, the beekeeper may be able to spin out some more honey towards the end of St. John's Tide (mid-July).

The honey harvest is always a festival for beekeepers and their families.

116

Figure 48. Capped honeycomb is opened with an uncapping fork

Figure 49. A four-frame tangential honey extractor

17. Summer Inspections

Because of the inner life rhythms of the honeybee, beekeepers always aim to inspect their colonies every nine days from the beginning of May to the end of June. Beekeepers are familiar with the microclimate at the site of their apiaries and keep an eye on how the weather develops. They also have a certain amount of knowledge concerning the vegetation and its stage of development within flying range of their hives.

It is necessary to get a feel for the mood that prevails in the surroundings of an apiary: the sights and sounds, the quality of the light and warmth, the condition of trees and the status of the vegetation. Activity around the hive is observed: the flight traffic, its direction, departures and arrivals, whether hunting about or direct, hesitating or hurrying, a little or a lot. If it is warm enough, then drones will be flying back and forth buzzing loudly from midday onwards. The orientation flight of the young bees shows that brood hatched in the colonies eight to ten days previously.

The ground in front of the hives is carefully inspected too. A significant number of dead bees, or small, deformed, wingless ejected brood, or bees incapable of flying (known as crawlers and hoppers), gives an indication of how healthy the colonies are.

Has there been a poisoning because a farmer or gardener has misused insecticide on open flowers? If so, there will be piles of dead bees in front of the hives and little further activity in the poisoned colonies.

Some of the incoming bees are carrying fat, colourful pollen baskets; yellow, orange and bright green catch our eye. Several bees hardly manage to complete the return flight and they land on the front of the hive above the entrance. After a pause they fly on again to disappear through the hive entrance or crawl head downwards down the wall into the entrance. Those bees returning without pollen are nectar gatherers or water carriers. A few bees are moving about on the alighting board facing outwards. They try to come into contact with incoming bees, briefly touching them with their forelegs and antennae. Hardly a bee enters the hive without being checked.

At times there are bees to be seen around the entrance and just inside it, their abdomens bent and pointing outwards as they hold firmly onto the floor and fan their wings. The easily felt stream of air that they drive out of the hive is warm and humid and smells sweetly aromatic, sometimes a little musty.

This behaviour tells the beekeeper that the bees have carried in a lot of nectar, which is being processed into honey by the bees in the hive.

Beekeepers know the history and characteristics of their bees, and thus a few careful observations of their behaviour and activity around some of the hives is enough to gain a good impression of how they are. Before the hives are opened one by one, the beekeeper sets out the necessary tools. These include a hive tool, a brush of some description, and a smoker to drive the bees away. The hive tool is a simple piece of metal with a chisel edge and usually a hook employed for a multitude of purposes such as scraping, shaving, levering, pushing and pulling. The brush can be a goose feather or a flat hand brush with a single row of soft bristles for sweeping bees off things when necessary.

The smoker is lit. Its fuel may be egg-tray cardboard, dried rotten wood, foliage or weeds (for example, tansy). In other

Figure 50. Beekeeping tools

words, any plant material that will smoulder well and for a sufficient length of time and which, at a squeeze of the bellows, produces smoke on demand that should not smell unpleasant.

After the roof and crown board and any insulation have been removed from the top of the hive, there may still be a plastic sheet or quilt covering underneath. This can be checked to gauge the warmth coming from the bees below. The more powerful the colony, the further the warm patch extends from the middle to the edges. It feels like body heat.

This last cover is removed. A sticky, reddish-brown substance glues it to the frames. This is propolis: a sticky resin that the bees use to fill up all the cracks in the hive that they cannot crawl through themselves. All inner surfaces are thinly plastered with it and polished. The bees collect the resinous, waxy coating from leaves and flower buds and process it into propolis by adding their own secretions and beeswax.

A few gentle puffs of smoke over the frames drives the bees

downwards into the gaps between the frames. The light-coloured surfaces of wax cappings can now be seen, which shows that the bees have already begun to produce and store honey. The frames are pushed together a little from the side in order to enlarge the gap at the side of the brood box. This allows the comb to be lifted out without squashing the bees that are closely packed together on it.

The comb has a beautiful ring of honey with white cappings at the top. Below this is a large area of cells filled with a firmly compressed substance in a variety of colours, sometimes coated with a gleaming film. This is the pollen layer of the brood nest, a comb whose cells are filled with flower pollen collected from a variety of sources and covered with a film of honey to preserve it. This is the bee bread.

The comb is carefully placed to one side and now the outermost comb is removed. A large part of it is already capped; it is almost completely filled with honey. On the next comb, under a ring of honey and pollen, there are cells of brood filled with round larvae of various sizes. The queen laid here between five and seven days ago.

Everywhere on the comb bees are busy cleaning cells, feeding brood, processing honey or feeding each other. A few foragers are also visible, dancing to stimulate other bees to fly to the forage source they have found. The lower margin of the comb has the larger cell pattern of drone brood. There are fat round grubs in them too. On the next comb in the lower part of the frame, we notice a large, semi-circular, capped patch of cells. This is the capped drone brood. On top of that is a patch of light brown, smoothly capped cells of worker brood, and under a ring of honey and pollen is a strip of as yet uncapped brood cells with fat round larvae in them. On the edge of the comb, in a small hole by the wooden frame, are some small semi-circular cones. Their openings point downwards and they have a closed-in appearance. These are queen cups in the making, and more can be seen when the next frames are inspected. Two empty frames inserted at the last inspection are already built in with beautiful white comb moulded with worker cells during a 'flow'. They contain young larvae as well as eggs.

Findings such as these express the vitality of a colony that is keeping its options open. The beekeeper may think about expanding the colony with a super to give it space to store honey.

The beekeeper does not inspect every colony in such detail. In some, the frames that have been added for the colony to expand are checked. The colony is only inspected more closely if these initial observations give rise to concern: for example, if the colony is no longer building comb, or if the comb is not in a single piece but has been started in several places at once and therefore appears out of harmony.

Swarm mood

In one of the stronger colonies, the super that was put on at the end of April is almost filled with fresh nectar and honey. Part of it is already capped.

If the comb is tilted, a few drops of a bright, clear liquid fall out that are sweet and aromatic to the taste. Fresh nectar is still rather runny.

The super is removed and carefully put to one side. A mass of bees boils up from the gaps between the combs. Inspection reveals large areas of honey and pollen and a sizeable brood nest with brood at all stages: eggs, young and old larvae, and capped brood. There is also a lot of drone brood and drones can be seen moving around on the margins of the combs. There are many cells with young bees just hatching from them.

It seems that every cell on every comb contains either stores or brood, and as a cell is vacated it is immediately cleaned and laid in again. The frames recently given to allow expansion are now completely built out with comb. On the margin of the comb are some queen cups but this time they look opened out, and on the polished floor of each is an egg. More queen cells are discovered. In them is a whitish, shining substance with a small round larva in it. The greater the length to which the cell has been drawn out, the larger the larva. One of the cells is even longer than the others and its opening has probably just been capped. A queen larva has a developmental period of three days as an egg and five days as a generously fed larva, and is capped eight days after egg-laying. Beekeepers know from this that on the next day a prime swarm will issue from the hive. The queen leaves the hive with a proportion of the bees and they swarm. The crops of the workers are filled with honey, otherwise the swarm will be taking nothing with it into an uncertain future. A week later the first young queen will hatch. Whether secondary swarms or casts will issue from the colony is determined by the bees and depends on the circumstances inside and outside the hive at the time.

A vigorous colony in swarming mood may easily raise young queens in a couple of dozen swarm cells. As these hatch, several casts can issue from the hive in succession. Some of these may contain several young queens. If the beekeeper catches such a swarm and later hives it, they may discover one or two dead queens on the floor of the swarm box. Obviously, the swarm has killed them and thrown them out. Either that, or the queens have fought one another and the winner has survived.

Swarm control

Swarming is a time in which beekeepers can rejuvenate, select and breed from their colonies.

If a beekeeper does not want to let the swarm issue due to the uncertainty of being able to find it and take it afterwards, then this can be prevented in various ways. First the queen is removed and hidden away with a few bees in a small queen cage.

Then the bees from at least three quarters of the brood combs are shaken into a swarm-catching box fitted with a large funnel. This yields an artificial swarm weight of 1.5–2 kg (3½ – 4½ lbs). The queen is then released from her cage into the box, which may now be placed in a cellar.

An artificial prime swarm must be fed in the cellar within 24 hours at the latest as the bees will not have been able to take as much of their stores with them as a natural swarm. The feed consists preferably of candied honey, although sugar syrup or fondant may be used if necessary. In addition, the composition of the bees is different, indeed arbitrary, according to which combs the beekeeper has shaken out into the artificial swarm. With a natural swarm, the bees that fly out of the colony have prepared for this by gradually filling their crops, amongst other things. Thus, a natural swarm can live for many days on the honey that it has taken with it.

A beekeeper can also prevent further casts from issuing. Instead of waiting for all the young queens to hatch naturally, the beekeeper can carefully cut the whole capped swarm cell out of the comb and let them develop in queen cages. Artificial casts are created and a queen cell, or a hatched queen, is given to each. These artificial swarms also have to be fed. They are hived in brood boxes with empty frames which the swarm fill with comb. Space in the box is usually restricted at first, either with a dividing board or a by using a smaller box of three or more frames. The swarm occupies this restricted space more readily and thus manages it better. This benefits the whole of its further development. The beekeeper then gradually increases the space with additional frames that are in turn filled with comb by the bees. If outer conditions are unfavourable it may be necessary to feed the swarms so that their development suffers no setbacks.

The young colonies develop in a similar way to that described in Parts II and III.

The beekeeper will maintain a close watch on these swarms to check how they are getting on and how well they manage to balance the important relationships between colony strength, comb construction, quantity of brood and increasing stores. In doing so, the beekeeper learns much that will be useful in later dealings with these colonies.

The pattern of reproduction and rejuvenation described above is usually carried out with colonies that, from their previous history over several generations of queens (queen lineages), the beekeeper regards as having good traits. Attention is paid to their health, vitality, strength, comb-building capacity, hygienic behaviour, eagerness to forage, honey yield, store development and management, docility, tolerance to being worked, and a number of other relevant characteristics, which may be important in the interaction between the beekeeper, the public and the bees.

Figure 51. Smoker

V. GIFTS FROM THE BEES

The Gift from the Bees

We work together all the day.
We work for others, not for pay;
So copy us if learn you will.
The sacred light from votive spill
Came from our crushed and melted comb
So think of this in your own home.
It should be said with great respect:
There is no work that we neglect.
The wax from off our abdomen
Perfumes the homes of bees and men.
We fashion cells in hanging tower,
Six-sided like the lily flower,
And store therein our flow'ry forage
That serves as bacon, eggs and porridge.
Their honey strengthens them and me:
So God, look kindly on the bee!

*Traditional verse from heathland beekeeping,
translated from German by Chris Slade*

18. Honey

When most people think of honey and of eating it, they associate it with a feeling of security, comfort, and abundance: bread and honey at breakfast on the morning of a holiday; the warm milk and honey that calms our flitting thoughts after the daily hubbub, or soothes a hoarse throat and softens the pain.

The aura of honey is strong and our ideas about honey are deeply connected with an inner sense of a mysterious force at work. The taste of honey embodies the joyful, amazing, confident mood of a midsummer morning, full of youthful freshness. Eating a little honey each day strengthens us for the day ahead.

Is it any wonder that Germans are the biggest consumers of honey in the world? In 2005 the average German citizen consumed on average 1–1.5 kg (2 ½– 3 ½ lbs) of honey.

Where does the wonderful power of honey come from?

Gathering nectar

Every child knows that when they see a bee on a flower it is collecting honey. But fewer and fewer people have really experienced the mystery of the little 'honey maker' (formerly Apis mellifica, now called Apis mellifera – 'honey bearer') or even looked into it in detail. But observing the process carefully enables us to learn something of the production of honey and ultimately of its significance for bees and people. It is not obvious that flowers contain a small amount of sweet fluid that is as clear as water. Only when the flowers are pulled apart do we really comprehend what bees are doing when they penetrate deeply into them. If we watch bees active in various flowers. we will realise that their body shape is optimal for such activity. In order to discover that bee abdomens contain a crop in which nectar is collected and which is big enough to carry a significant amount, we first have to dissect them (see Figure 33 on page 66). And it takes many more observations to show that individual foraging bees are in no way directly self-motivated in their activity.

A bee on her foraging flight usually visits flowers of the same kind – for example, apple blossom – in a small area until her crop is full. Then she flies back to the hive. There she transfers the nectar to the house bees and soon flies off again to the same kind of flowers in the vicinity as before, as long as they are yielding nectar. Thus, honey bees are constant in their preference for a particular flower species. Only when a source runs out – that is, when the secretion of nectar by the flowers visited so far decreases – are the bees then stimulated by the dances of other foragers in the hive to fly to a new, more attractive source of forage.

In the course of the year, the foraging zone of a particular colony changes. In spring the flight radius remains rather short, encompassing the main nectar sources such as fruit blossom, dandelion, rape and, later, woodland forage or that of arable flowering crops. The radius soon increases to 1.5 km (just less than 1 mile), which corresponds to an area of 7 km^2 (700 hectares), or at times even to 3 km (just less than 2 miles), almost 30 km^2 (3,000 hectares) or more. The flight radius is dependent on the strength of the colony, weather conditions and the attractiveness of a particular source of forage. The latter is predominantly dependent upon the flower density, the nectar yield of individual flowers, the sugar content of the nectar and whether this is easy to get at in the flower. The more favourable these factors, the more bees fly to particular sources of forage even if they are further away.

The nectars of various flowers differ in sugar composition, aroma and other constituents. Foragers of a colony kept permanently in one location generally visit most of the species that flower in their flight range as long as they yield nectar and/or pollen. As a result, there is a constant mix of nectars coming into the hive – more from one species, less from another – depending on what is available. The variety of nectars is dependent on the diversity of flowering plants within flight range.

As a rule, honey made from specific flowers (for example, fruit blossom, rape, chestnut, sunflower, etc.) can only be harvested by beekeepers who place their colonies in large expanses of a particular source of forage. Beekeepers move their bees to orchards, rape and sunflower fields, sweet chestnut blossom or other large honey-producing stands, as well as deciduous woodland, pine areas and heather moors, soon after the flow begins at them. Migratory beekeeping also occurs; that is, beekeepers move with their bees.

Forager bees reorienting at an unfamiliar site will find these plants first and subsequently the bulk of the bees will visit only these.

How bees turn nectar into honey

The nectar brought to the hive must undergo special processing before it is changed into honey. Whereas nectar is usually a slightly sweet fluid as clear as water, honey has a viscous to solid consistency and is usually very sweet and aromatic. The smell and taste may be slightly sweet or very fruity, tart or even bitter-sweet. The water content of nectar is up to 80%, whereas the residual moisture of honey is only 15–20%. Nectar contains a variety of sugars, several of which cannot be used as sources of energy, either for bees or for people, because they cannot be digested in the form in which they occur. The sugars in honey, and here the term 'sugar' is being used in the same sense as in analytical chemistry, are primarily fructose and glucose, plus a small amount of other types of sugar. The colours of various honeys may range across the whole palette of earth colours from almost pure white, via light yellow, deep yellow, beige, ochre, golden brown, brown, dark brown to almost black.

Once a forager has taken up sufficient nectar from the flowers, she commences the return journey to the hive. Whilst collecting she visits 100–200 or more individual flowers, enriching the nectar with secretions from her royal jelly and salivary glands. These secretions are extremely rich in various enzymes, active substances which catalyse chemical reactions in living organisms. In this case they bring about changes in the indigestible nectar sugars, which turns them into the utilisable sugars, fructose and glucose. The enzymes also play a significant part in the beneficial effect of honey for people.

After their foraging flight, the foragers transfer the drops of nectar to the house bees. Subsequently, these drops are passed on many times from bee to bee and are further enriched by secretions from their glands. The more bees that participate in this processing chain the more concentrated the resulting honey. The nectar undergoes a significant change during the evaporation of its water content. It is actively concentrated by the bees. A bee engaged in honey processing ejects a droplet from its crop, spreads it out flat on its outstretched proboscis and then draws it together again. It repeats this procedure rhythmically every few seconds for up to fifteen to twenty minutes. While this happens further glandular secretions are added. The now partially ripe honey is then spread in a thin layer in empty cells in the warmth of the brood nest area, or deposited in tiny droplets on the cell walls.

Returning forager bees transfer their loads to the young house bees

The young bees expose the nectar to the air by regurgitating it and taking it in again. This evaporates the water and as it passes the glands in the mouth substances important for the transformation of nectar into honey are added.

Stages in the process of regurgitation and taking in. The proboscis is slowly extended and swung back again.

1. Proboscis closed
2. A drop of nectar pumped back out of the crop appears in the glossal groove
3. The drop increases in size and partly flows downwards between the stipes and the galea
4. Further increase in the amount of nectar
5. Larger drop of nectar, held by the increasingly opened galea
6. Rhythmic opening and shutting of the galea in which the drop is repeatedly consumed and pumped out of the crop.

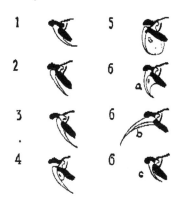

Figure 52. How bees turn nectar into honey

The legal definition of honey (according to German honey regulations)

Fluid, viscous fluid or crystalline foodstuff produced by bees by collecting flower nectar, other secretions of living parts of plants or secretions on living plants from insects, enriched and modified by the secretions of the bee's body, deposited in comb and ripened there.

Sugar compounds	The average content of fructose is 38% and of glucose 31%. The content of maltose and other disaccharides and oligosaccharides is c. 9%
Vitamins	Honey contains traces of vitamins B1, B2, B6, C, biotin H, pantothenic acid, nicotinic acid and folic acid. It lacks the fat soluble vitamins.
Minerals	Potassium, magnesium, calcium, silicic acid, phosphorus, sulphur, manganese, silicon, sodium, copper, chlorine
Trace elements	Iron, copper, manganese, chromium, etc.
Amino acids	Glutamic acid, leucine/isoleucine, aspartic acid, phenylalanine, threonine, alanine, histidine, glycine, lysine, serine, valine, proline, cystine
Enzymes	Invertase, amylase, catalase, phosphatase, glucose oxidase
Hormones	Acetylcholine, growth hormones
Acids	malic acid, acetic acid, citric acid, lactic acid, butyric acid, succinic acid, pyroglutamic acid, gluconic acid, hydrochloric acid, phosphoric acid, formic acid.
Pollen	
Aromatic substances	Organic acids, phenylacetic acid ester, acetaldehyde, isobutyraldehyde, formaldehyde, acetone, diacetyl and a further 120 fragrance and aromatic compounds.

Table 3. Honey ingredients

In the dry warm air-stream (35°C/95°F), 40–50% relative humidity) that flows over the comb surface, more water is evaporated from the honey until, after between 1 and 3 days, it falls to 20% residual moisture. The bees then carry it to the margins of the comb, usually on the opposite side from the hive entrance. Once the cells there are filled with ripe honey (less than 18% residual moisture) they are closed with an airtight wax capping. At the honey harvest the beekeeper can often find the same forage deposited in adjacent cells in patches of varying size. This shows the extent of the constancy of the bees' flower source.

In considering the overall phenomenon of honey production, two qualities can be experienced. The quality of the flowering plants in the environs of the hive is present in the honey in a coagulated, concentrated form. The substances are brought to a point in this way, not only spatially in the sense of flower diversity, but also temporally according to the sequence of flowering.

As a result, honey is a reflection of the relationship of each colony with its surroundings. Furthermore, it is an expression of the ability of the hive organism to manage the incoming nectar and to enrich it with its own secretions during the ripening activities. Thus, the honey from a number of neighbouring colonies harvested by the beekeeper on the same day, but extracted, ripened and bottled separately, is likely to differ in appearance, aroma and consistency.

Returning forager bees transfer their loads to the young house bees.

The young bees expose the nectar to the air by regurgitating it and taking it in again. This evaporates the water and as it passes the glands in the mouth, substances important for the transformation of nectar into honey are added.

Stages in the process of regurgitation and taking in. The proboscis is slowly extended and swung back again.

19. Pollen, Bee Bread and Royal Jelly

The value of pollen for bees

The process of gathering pollen is more visible than that of nectar and can be observed with the naked eye. Whereas pollen gatherers work more superficially on the flowers, nectar gatherers often have to force their way deep inside the flowers in order to reach the nectaries with their tongues. They frequently have to burrow, roll, dance right into the chalice of the flower as they manoeuvre, rip open and shake out the anthers with their mandibles and forelegs. The whole bee body is engaged in filling the pollen baskets and we soon notice the remaining pollen caught on the hairs all over the head, thorax and abdomen. At the same time, the free pollen is constantly wetted with fluid from the proboscis, either from the honey the bee has brought with her in her crop, or from the nectar extracted from the flower. This helps to stick the pollen together.

As soon as the bee flies out of the flower, she cleans herself with her three pairs of legs and in doing so forms the pollen baskets. The head is cleaned with the combs on the forelegs, the thorax with the middle pair. These also pick up the pollen collected by the forelegs. The forelegs are then scraped on the tarsal combs of the hind legs. Furthermore, these are rubbed together in parallel so that the pollen clumps are scraped out of the combs of one hind leg into the pollen rake of the other in a reciprocal process.

Push by push the pollen baskets are filled. When this process is almost finished, the pollen is packed down firmly by the middle legs, during which fluid is again pressed into the upper layer of pollen. These movements take place so quickly that we cannot easily follow them with the eye. But frequent observations over a short period of time enables a complete picture of the process to be built up (see Figures 5–7, page 15).

The pollen baskets vary in size and weight depending on the source and yield of the flowers. The average pollen load (two pollen baskets) weighs 8–12 mg and sometimes reaches 18–20 mg depending on the pollen available and its composition. The pollen colour differs between plant species too and covers the entire colour spectrum, but because bees show constancy of flower preference in pollen gathering (see page 130) only rarely are the baskets multicoloured.

Returning bees scrape the pollen baskets into a comb cell by holding onto its upper margin, allowing their rear legs to hang deep into the cell, and then freeing the lumps of pollen from

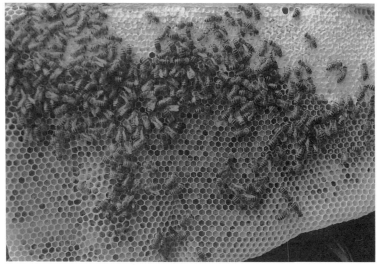

Figure 53. Store comb; capped honey at the top, bee bread below it

137

the basket surfaces. The pollen then falls to the floor of the cell. But here the constancy of flower preference stops. The pollen in the cells is generally of mixed origin as well as a mixture of colours.

The house bees who are responsible for this task further moisten the pollen mass and knead it. They stamp it into layers in the cell with their heads until it is about three-quarters full. The mass is then covered by a layer of honey and stays like that until it is needed. In this condition the pollen undergoes a kind of mild lactic acid fermentation, which preserves it to a limited extent and breaks it down for the bees to digest. The pollen stored in the comb in this way becomes what is called 'bee bread' or 'perga'.

In scientific experiments, freshly hatched bees were isolated and put with a queen in an artificial colony. They received only honey or sugar solution and pollen from a pollen ageing experiment. The pollen spoiled very quickly and lost its nutritional value. As a result, the bees showed only limited activity and did not develop sufficient glandular secretions to care properly for the brood. Further development of individual bees did not proceed as normal.

From this somewhat crude negative evidence, we can see that pollen provides the nutritional basis for forming the bodies and structure within the colony. By consuming pollen, individual bees acquire the impulse to develop their intrinsic nature and perform the necessary jobs in the hive in a certain orderly sequence.

Pollen and people

Consumers can buy pollen baskets mixed in colours in jars, or in packets, either in shops, or directly from beekeepers. Beekeepers harvest them by fitting a pollen trap to the entrance of the hive. This is a plate with a small, round or star-shaped opening. The pollen gatherers must force themselves through these openings which causes them to lose some of their pollen baskets. The beekeeper then cleans and dries the pollen, which would otherwise quickly spoil, so that it can be offered for sale. This flower pollen, in contrast to honey, has not yet undergone any development within the hive and is thus still to be regarded as raw material. Nutritional physiologists are quite unable to agree about whether human digestion can make significant use of this form of flower pollen: the individual pollen grains are covered with a tough outer skin, which is not easily broken down by the digestive juices.

It is different with the pollen in the comb, with the bee bread. Here the fermentation process has already opened the pollen skin so that all the valuable constituents of pollen can be digested. But this bee bread is rarely stored to excess in colonies, and it keeps only for a limited period of time. Thus, it is most unusual that a beekeeper can harvest any of it, and it is a special blessing if we are ever personally presented with this bee gift by a beekeeper.

Royal jelly

Royal jelly can be purchased for a substantial sum from health food shops. It is a secretion of special glands in the bee head and is produced by young bees at a particular stage of their development, provided that they consume sufficient quantities of honey and bee bread. The name 'royal jelly' arises from the white, viscous, gelatinous plug in which the queen larva hangs. The nurse bees feed the larva in a queen cell so generously with royal jelly that after the hatching of the 'princess' a lot remains dried up in the cell. The royal jelly that the workers and drones are fed with is clearer and less opaque than queen royal jelly.

Although the productivity of nurse bees is distributed over several brood cells, on average a worker bee may raise the equivalent of two larvae with the royal jelly she produces. This shows the great metabolic output of the nurse bees. They create a completely new substance out of honey (which from the point of view of nutritional physiology works primarily as an energy supplier) and out of bee bread (which supplies protein, fat, starch, vitamins and mineral substances).

It is no wonder that people are so interested in royal jelly and invest it with energy-giving, health-promoting and life-prolonging properties. Indeed, a queen who is fed exclusively with royal jelly can sometimes live for as long as five years, and produce many hundreds of thousands of eggs. In contrast, in summer, a worker that is fed for only three days of its larval stage with royal jelly and then a further three days with honey and bee bread, may live for a maximum of only six weeks, and in winter perhaps reach an age of nine months. Thus, scientists are interested in finding out the as yet undiscovered constituents and properties of royal jelly.

But one thing is clear, the royal jelly can only be found right in the middle of the bee colony, and where honey and bee bread are still recognisably similar to their floral origins, royal jelly is a pure creation of the bee organism.

20. The Power of Honey for Bees and People

Summer honey for bees

It should first be made clear that a lot of honey is given as food in the raising of bee brood, whether directly in feeding the older brood, or in the form of royal jelly for nursing the young brood and the queen. But the value of honey to bees can be further differentiated as illustrated in the following example.

If a bee has flown inside a house, and lies exhausted on the windowsill after long efforts to get out again, it will soon show no signs of life. But if you bring a drop of honey on your finger, the bee will soon begin to lick it up. After a few seconds, it will look more lively and begin to pump its thoracic and wing musculature. Its wings will stretch out, tremble a little. When the bee has consumed sufficient honey, it will crawl away and finally fly off. This example shows that honey provides the energy for the outer activity of bees, such as the activities we observe during foraging.

But the inside of the hive presents another picture: the bees release the necessary heat for the healthy development of the brood by working their thoracic and wing musculature while staying in one spot. The ability to regulate the temperature of the colony is so strong that the necessary temperature of 35.5 °C (96 °F) can be maintained in the brood area even when it is extremely cold outside. This has been observed in freezer room experiments at temperatures as low as -80°C (-112°F), although this led to an increased consumption of honey. Likewise, the hive is ventilated by the fanning of bee wings near the hive entrance and, if it is very hot, water is evaporated in the hive to cool it. The building of comb with wax exuded by the bees also requires a plentiful supply of energy from honey. Thus, in beeswax, another completely new substance is created in the hive which has little in common with the substances used in its production.

If we bring these facts together into an inner picture, we can see that honey is a prerequisite not only for the outwardly visible and observable activities of bees, but also as the basis of their physical life and the manifestation of their nature through the forming of the colony.

Winter honey for bees

Honey is also important to bees in its storage role. In Central European latitudes (46°–57°N), the period in which bees can accumulate a surplus out of the surroundings of the hive (from the end of April to the beginning of August), makes up only a few whole days in many years. The rest of the time they are usually only able to bring in enough for their immediate use, and in bad weather they have to consume their stores. The beekeeper harvests part of the surplus and the remainder must meet the requirements of the bees when they can no longer bring anything in from outside. Bees, as we have seen, do not hibernate but instead over-winter with a certain amount of outward inactivity. But they do not do so as individual animals, as do bumblebees, wasps or hornets, the queens of which enter into a winter paralysis in a hiding place somewhere. Honeybees pass the winter months together as an organism in the hive. They are awake but only active within themselves. The reason why they store honey in significant amounts at all is that they need it to survive the winter. Their stores may be more than they need, in which case we can harvest some of them.

Honey is the result of the sense perception of the bee colony in summer: smelling and tasting whilst visiting flowers; smelling, tasting and feeling during the processing of honey from nectar gathered in the hive. This sense perception is deposited in the form of honey in the combs. The bees consume it in winter and this reactivates the sense perception inside them. People refer to sense perception that is reactivated within as 'memory'. Likewise, we could say that honey is itself memory. For the bees in winter it is a memory of their activity in the heat of summer, an intimation, even a foretaste, of the hot summer to come. This can be summarised in the following way: honey keeps the bees in the physical world.[4]

We have discussed the many manipulations that the nectar must undergo before it is stored as ripe honey by bees in the cells of the comb (see page 133). In this respect it should also be made clear that although the process of ripening nectar into the honey we consume takes place in the hive, it does not go through the metabolism of individual bees, nor become a new substance. The honey ripens within the hive, but outside the individual bees. It is concentrated and purified with the bodily secretions of the worker bees, rhythmically moved and looked after by them. We can say it is potentised by them.

Rudolf Steiner on honey

In his lectures on bees to the workers at the Goetheanum in Dornach, Switzerland, the philosopher Rudolf Steiner said that bees are creatures that live primarily in smelling and tasting. If we study bees closely we can confirm this statement. The sense of sight is of secondary importance to bees. When a bee sets off foraging it does not search with its eyes. Stimulated by the dance of the other bees in the hive, the foragers know when they are flying off course and where they are heading (see Chapter 7 for a description of the bee dance). They do not have to search. They find their way to the forage site through an inner sense and once there they follow their sense of smell. They taste whether or not they are on the right flower, and smell and taste the nectar as they gather it. The house bees smell and taste the nectar too as they process it, adding their secretions and concentrating the honey until they taste that it has indeed

Figure 54. Swarm cluster

Figure 55. Swarm cluster from below

become honey. The production of honey is therefore the result of the smelling and tasting of the whole hive.

Rudolf Steiner indicated that consuming honey strengthens the human I-organism (ego). This statement can be fully explained through studying the phenomenon of honey preparation in connection with the special sensory organisation of bees. The human I-organism needs to assert itself constantly against the impressions of the senses. Honey, a result of sense processes, is a medium that helps the I-organism assert itself against the flood of sensory stimuli, so that people do not function in the world solely according to their instinctive reactions.

In a lecture on 22nd December, 1923, again to the workers at the Goetheanum, Rudolf Steiner described a particular aspect of the relationship between the human being and a swarm of bees:

> As I told you, when a young queen hatches in the hive there is something now amongst them which disturbs them. Previously the bees had lived in a kind of twilight. Now they see the young queen begin to shine. What is connected with this shining? It is connected with the fact that the young queen takes over from the old queen the power of the bee venom. And what the departing swarm fears, gentlemen, is that it will no longer have the bee venom. The bees are afraid that they will no longer be able to protect or save themselves. It departs just as the human soul does at death when it can no longer produce formic acid. The older bees depart when insufficient transformed formic acid, that is, bee venom, is present in the hive. And when we now look at the swarm, although it is of course visible, it looks like the human soul when it has to leave the body. An issuing swarm is a magnificent sight. Just as the human soul leaves the body, so also the old queen leaves the hive with her followers when the young queen is ready. And in

145

the issuing swarm we can see a real picture of the departing human soul.

Gentlemen, this is truly wonderful! But the power of the human soul has never managed to go as far as making itself into little creatures. Nevertheless, there is always a tendency in this direction in us; always wanting to become like little creatures. We really have in us this constant desire inwardly to turn ourselves into crawling bacilli and bacteria, into tiny bees, yet we suppress it so that we can be fully human. But a hive of bees is not a complete human being. The swarming bees cannot find their way into the spiritual world. It is we who have to reincarnate them in a new beehive. This is a direct picture of human reincarnation. And anyone who has the opportunity to observe such a thing has an enormous respect for these swarming old bees with their queen who behaves as she does because she wants to go into the spiritual world. But she has become so physically material that she is unable to do this. And then the bees huddle together; become a single body. They want to be together. They want to be out of this world. As you know of course: whereas they otherwise would be flying, now they settle on the branch of a tree or on something else, snuggle up to one another, try to disappear because they want to go into the spiritual world. And then they become a proper colony once again when we help them by returning them to a new hive.[5]

Harvesting honey

When the comb is ripe it is harvested by the beekeeper and honey is extracted by spinning or pressing. The honey is ripened artificially by filtering and stirring and is then left to stand before being bottled to adorn our breakfast tables. 'Comb honey' is a particular delicacy. It is honey in its purest form in its original packing of hexagonal cells. The comb, freshly built by the bees, is a brilliant white and has not had any brood in it. It is comparable to delicious confectionary.

Storing honey

Bees process honey in the darkness of the hive at a temperature of about 35°C (95°F). It is somewhat cooler on the margins of the comb where ripe honey is stored in capped, airtight cells. It is only warmed once more when the bees take it up into their organism again.

After the beekeeper has extracted the honey it cools down. It is then processed further and stored in airtight jars, usually before its first setting. The honey should be kept in a cool, dark place, not in a refrigerator, and then it will keep its quality for an almost unlimited period of time. Light and warmth activate the enzymes in honey in much the same way that happens when it is taken up into the human organism. It is for this reason that honey has beneficial properties. If it is kept in the light and heat it steadily loses its value. When it is heated to make it more fluid, the enzymes are activated and it changes, the honey is altered and it loses its effectiveness.

What should honey really cost?

In the same set of lectures on bees to the workers at the Goetheanum, Rudolf Steiner said on the subject of honey: 'Honey is something so valuable that it is impossible to put a price on it.'

A small calculation

On the foraging flight, which lasts 30–45 minutes, a bee visits 200–300 flowers of a plant species. In doing so it accumulates in its crop about 0.05 g of nectar, roughly half its body weight. In really good weather a bee will manage about ten sorties a day which yields 0.5 g of nectar. On a good day, 10,000 foraging bees on 100,000 foraging flights will work for 50,000–75,000 hours visiting

20–30 million flowers and in doing so may bring in around 5 kg (11lbs) of nectar. This is processed into 1.5 kg (3.3 lbs) of honey in the hive, approximately 4 standard jars. If bees were paid a living wage of £8 each per hour, a jar of honey should cost £100,000 for the nectar gathering process alone.

Conversation with a queen bee

'If I politely, kindly, ask
For honey in a jar or flask,

What would you charge me? How much please?
I can afford the best with ease.'

'You want a product pure and sweet?
You want the best that you can eat!

You're buying sunshine. Summer's through.
You know that honey's good for you!

There's nothing better to be found.
I'll calculate the price around

The work my bees did at the flowers.
It took them twenty thousand hours.

It's easy in the honey flow
And, yes, their pay is very low.

A pound an hour, or mark, or yen;
That's twenty thousand of them then.

With twenty thousand pounds or more
I'm glad you're rich while we are poor!'[7]

Josef Guggenmos, from Was denkt die Maus am Donnerstag?
English version by Chris Slade.

Appendix: Demeter Beekeeping
by Günter Friedmann

A new approach to organic apiculture

In the summer of 1995 the Demeter Association introduced guidelines for the production of honey under Demeter certification. They were developed in collaboration with the Demeter Association's specialist subgroup on beekeeping.

The Demeter Association is the oldest organic farming organisation in the world. In the UK, Demeter certification is managed by the Biodynamic Agricultural Association (www.biodynamic.org.uk). Its member producers, processors and traders are interested in doing more than just working close to nature without artificial fertilisers and plant protectants. Indeed, a special aim of this farming association is the care and development of soil fertility for which certain biodynamic preparations are used. Demeter principles and practice seek to put *culture* back into agri*culture*.

The Demeter guidelines for beekeeping have the same aims. The concern is not just the avoidance of residues in honey, wax and propolis, but keeping bees in accordance with their nature. This is the only way to ensure the productivity and vitality of bee colonies in the long term.

This aim is addressed explicitly in the foreword to the guidelines:

> Beekeepers work in the context of biodynamics and orientate themselves primarily towards meeting the natural requirements of the colony. Management is so structured that the bee is able freely to unfold its true nature. Demeter beekeepers allow the colonies to build natural honeycomb. The basis for their reproduction, growth, rejuvenation and breeding is the process of swarming. Its own honey is the mainstay for supporting the colony through the winter.

Demeter beekeeping in no way requires that bees forage only over land that is under organic management. The crucial thing about Demeter beekeeping is how the bees are managed.

It would not be surprising if many beekeepers find such statements unacceptable. They challenge fundamental assumptions of so-called modern beekeeping, such as systematic swarm prevention, making nucs (nuclei), artificial queen breeding and insemination. Many beekeepers believe that successful and profitable beekeeping is possible only with such measures. Yet an increasing number of beekeepers who work according to the Demeter guidelines, whether they have only a handful of colonies or several hundred, have found another way of dealing with bees that can also be successful and profitable.

The guidelines assert that like agriculture there is a *culture* which is particularly for beekeeping, that is, *apiculture*. However, it does not mean that beekeepers should do nothing and leave colonies to do as they please. On the contrary, they have a good many possibilities for intervention. But such interventions are not measured against maximising honey yield but against the requirements of bee colonies themselves. Since healthy and populous colonies also bring in respectable amounts of honey, such an approach to beekeeping is also economically viable.

The Demeter guidelines focus on the way beekeepers manage their bees. Demeter beekeepers support a very consistent approach which avoids all systematic and arbitrary nuc production. Increase is only made via swarming, which is the natural method of increase and reproduction. This does not mean that beekeepers have to stand around doing nothing and just wait until swarms are hanging from trees or flying away.

The important thing is not adhering rigidly to using all strong colonies for making up nucs and queen breeding, but intervening only in colonies which are intent on increase and reproduction because they are showing the symptoms of imminent swarming. They may then express their natural drive and go through the swarming process to a certain point. From the practical angle this means, first of all, that beekeepers must carry out regular swarming checks. If they then establish that some colonies are in a swarming mood they must not attempt to suppress this urge but make artificial swarms with the old queens of such colonies. The rest of the original colony may then be left or divided into nucs with swarm cells.

According to Demeter beekeepers, queens are not queens in the full sense of the word unless they have developed naturally; that is, they must be swarm or supercedure queens. Raising emergency queens is what bee colonies do when faced with the risk of dying out and should not form the basis of increase and reproduction. Dispensing with artificial queen breeding does not mean doing without selection altogether. Beekeepers are still free to choose only cells from good colonies. But the aim of selection should be to keep bees of European races that are adapted to the landscape and region and to avoid crossing them with bees from other continents.

Comb management is also part of the procedure. Demeter

beekeepers regard comb construction as an integral part of a bee colony. The brood nest represents a natural unity. According to the guidelines, beekeepers should use only frames of a size that enables the brood nest to develop without hindrance from the frame bars (for example, Dadant size).

Apiculture that follows Demeter guidelines regards and treats the bee colony as a whole organism. Bees, brood, queen, stores, comb etc. all form a unity which should not be disrupted. Thus, in the brood nest region, the central region of the colony, all the comb should be constructed by the bees alone – natural comb building. The brood nest and the comb body, mediated by the natural comb construction, can then be appropriate for the size of the colony. Consequently, queen excluders are dispensed with.

Comb foundation may only be used in supers. This is a compromise between economy and ecology. Even so, the foundation used should be manufactured from natural comb. The reason for this requirement is connected with the quality of the wax. All kinds of residues accumulate in wax. Furthermore, wax deteriorates in the course of endless wax recycling. Naturally-made wax guarantees a not insignificant wax harvest and a wax of a high purity and quality. This finds an easy market as high-quality wax is the best base for natural cosmetics.

Honey is the basis of the natural nutrition of bee colonies. Consequently, Demeter beekeepers need to add a certain amount of honey to winter feeds (at least 10% by weight). With this, and through the addition of salt and herbs, the nutritional quality is increased.

Honey from Demeter beekeeping should be bottled for sale before the first setting. This avoids the reduction in quality which happens at temperatures even below 40°C (104°F). The requirements of these guidelines pose significant challenges to beekeepers, especially at the time of the flower honey harvest, but they can be met by modern bottling technology. However, honey consumers also need to be made aware that during a particular part of the year only set honey is available.

The higher quality of Demeter honey warrants the higher price which, supported by the good reputation of the Demeter certification mark, is easily realisable.

All organic beekeeping organisations take for granted that *Varroa* mites should be controlled by formic, oxalic and lactic acids and not by artificial chemicals such as pyrethroids.

A beekeeper requiring Demeter certification must normally go through a conversion period in which they are allowed to sell honey as being from apiaries 'in conversion to Demeter' beekeeping. The entry criteria are such that beekeepers who already practise beekeeping to a particular organic standard may enter more easily, and may be able to develop this very different approach to beekeeping to full Demeter certification status within three years.

No doubt the expectations of the Demeter guidelines are high and demanding for beekeepers. Demeter recognition does not come on the cheap. The extra effort is worthwhile because implementation of the guidelines in one's own apiaries leads to a way of managing bees that is of an inherently high organic quality.

About the author

Master Beekeeper, Günther Friedmann (b. 1956), is a professional beekeeper and manages his bees according to Demeter guidelines. He is a co-director of the Demeter Association in Germany and manages the organic production and processing standards inspections. His business is situated at Steinheim in the Swabian Alps and is a recognised training centre.

Acknowledgements

I would like to thank the pupils at Jean-Paul School in Kassel. In 1992, I worked with them on a Main Lesson book on the theme 'Apiology,' which became the initial basis of this approach for me. I am also grateful to the participants of many introductory courses on the theme of 'Man and Bee' for their patience and their questions, which helped me to develop further phenomenological presentations.

I received many useful suggestions as a result of my work with Illse Müller, Heidelberg, on 'Exercises and observations for approaching bee colonies' and from the expert subcommittee on beekeeping at the Demeter Association. Also, from Dr Jochen Bockemühl, and from the work of the Bee Study Group of the Science Section of the School of Spiritual Science in Dornach, Switzerland.

Over a period of many years, Michael Olbrich-Majer of the Research Group for Biodynamic Agriculture in Darmstadt devoted many pages of the journal *Lebendige Erde* to my work. And Uli Nett, Kassel, has been a partner in our beekeeping venture for fifteen years and a real friend to me in my work.

Tyll van der Voort of Oaklands Park Camphill Community had the initiative for the first courses on beekeeping in Britain. I want to thank him and Bernard Jarman of the Biodynamic Agricultural Association (BDAA) for their commitment and organisation of the courses. Bernard was also the driving force behind the translation of this book into English. David Heaf is a beekeeper, and this gave added depth to his translation. I would like to thank him especially for the excellent working relationship during the production of this book.

Thank you to everyone at Floris Books for publishing this translation.

I would also like to thank the participants of the courses I was able to give in Britain. They enabled my wife and I to get to know this wonderful country and allowed me to discover that I can speak English for three days.

Special thanks must go to the bees: their images in me were obviously so strong that they overcame language barriers.

Finally, and above all, I wish to thank my wife, Monika, who in my years of beekeeping has patiently borne the stickiness of honey and my lack of punctuality, and who took care of our lively children while I looked after my bee colonies.

End Notes and Picture Credits

1. From *Goethe's Botanical Writings*, Bertha Mueller, University of Hawaii Press, 1954.
2. Griziwa, J., Büdel and Herold, p. 17.
3. This chapter is based on 'Winter contemplations', produced in a Main Lesson on beekeeping in Class 5 of the Jean-Paul Waldorf School, Kassel, Germany, summer 1992.
4. *Frederick* by Leo Lionni illustrates this phenomenon of warming memories in a simple yet surprisingly profound manner.
5. *Bees*, 8 lectures, Dornach, 3rd February to 22nd December 1923, GA 351, trans. T. Braatz. New York: Anthroposophic Press.

Figures taken, with kind permission, from the following sources:
- Zander and Weiss: 1, 7, 13, 16, 20, 26–29, 32, 38–40, 42
- Büdel and Herold: 2–4
- Dorothy Hodges, 1952: 5–6
- Doering and Hornsmann: 8–10, 37, 41
- Tim Kraus, Kassel, 1995–6: 11, 18–19, 21–22, 30, 47–49, 51
- Author: 12, 14, 17, 25
- Ruttner: 15
- Klaus Bogen, Kassel, 1994: 23–24
- Casaulta, Krieg and Spiess: 31, 33–35
- Günther Friedmann, 1998–9: 36, 53
- Herold and Weiss: 43–46, 50
- Park: 52 (originally Park, also in Büdel and Herold and Casaulta, Krieg and Spiess)
- Rainer Vietor, Kassel, 1995: 54–55

Bibliography and Further Reading

Bresette-Mills, Jack, *Sensitive Beekeeping: Practicing Vulnerability and Nonviolence with your Backyard Beehive, Edinburgh*, Floris Books, 2016

Bruyn, Clive de, *Practical Beekeeping*, Crowood Press, 1997

Büdel, Anton & Herold, Edmund, *Bienen und Bienenzucht*, Munich, 1960

Casaulta, Glieci; Krieg, Josef; & Spiess, Walter (eds.), *Der Schweizerische Bienenvater*, Aarau, Switzerland, 1985

Davis, Celia, *The Honey Bee – Inside Out*, Bee Craft Limited, 2004

Doering, Harald & Hornsmann, Erich, *Die Welt der Biene*, Munich, 1956

Hauk, Gunther, *Toward Saving the Honeybee*, Edinburgh, Floris Books, 2017

Herold, Edmund & Weiss, Karl, *Neue Imkerschule*, Munich, 1995

Hooper, Ted, *Guide to Bees and Honey*, Marston House Publishers, 1998

Kornberger, Horst, *Global Hive: What the Bee Crisis Teaches Us About Building a Sustainable World*, Edinburgh, Floris Books, 2019

Lionni, Leo, *Frederick*, Alfred A Knopf (USA) and Penguin Random House (Canada), 1967

Maeterlinck, Maurice, *The Life of the Bee* (1901), Dover Publications, 2006

Mueller, Bertha, *Goethe's Botanical Writings*, University of Hawaii
    Press, 1954
Park, Wallace, *The storing and ripening of honey*, Rep. of the State
    Apiarists f. 1923, State of Iowa, 1924
Ruttner, Friedrich, *Naturgeschichte der Honigbienen*, Munich,
    1992
Steiner, Rudolf, *Bees*, 8 lectures (GA 351), Dornach, Trans. T.
    Braatz, New York, Anthroposophic Press, 1998
— *Harmony of the Creative Word* (GA 230), London, Rudolf
    Steiner Press, 2001
— *Agriculture Course: The Birth of the Biodynamic Method* (GA
    327), London, Rudolf Steiner Press, 2004
Winston, Mark L., *The Biology of the Honey Bee*, Harvard
    University Press, 1987
Zander, Enoch & Weiss, Karl, *Das Leben der Biene*, Stuttgart,
    1964

Useful Organisations

AUSTRALIA
Biodynamic Agricultural
Association
www.biodynamics.net.au

CANADA
Demeter Canada
www.demetercanada.ca

GERMANY
Demeter-Bund
www.demeter.de
www.demeter.net

IRELAND
Biodynamic Agricultural
Association of Ireland
www.biodynamicagriculture.ie

NEW ZEALAND
Biodynamic Association
www.biodynamic.org.nz

SOUTH AFRICA
Biodynamic Agricultural
Association of Southern Africa
www.bdaasa.org.za

UK
Biodynamic Association
www.biodynamic.org.uk

USA
Biodynamic Association
www.biodynamics.com

Michael Weiler
www.Imkerberatung.de
www.Der-Bienenfreund.de
www.Demeter-Imker.de

Floris
Books

For news on all our **latest books,**
and to receive **exclusive discounts,**
join our mailing list at:

florisbooks.co.uk

Plus subscribers get a FREE book
with every online order!